Thomas E. Welch

MOVING BEYOND ENVIRONMENTAL COMPLIANCE

A Handbook for Integrating Pollution Prevention with ISO 14000

LEWIS PUBLISHERS

Boca Raton Boston New York Washington, D.C. London

Library of Congress Cataloging-in-Publication Data

Welch, Thomas E.
    Moving beyond environmental compliance : a handbook for
integrating pollution prevention with ISO 14000 / Thomas E. Welch.
      p. cm.
    Includes bibliographical references (p.   ) and index.
    ISBN 1-56670-295-X (alk. paper)
    1. ISO 14000 Series Standards--Handbooks, manuals, etc.
2. Pollution prevention--Handbooks, manuals, etc. I. Title.
TS155.7.W45 1997
658.4′08--dc21

97-27474
CIP

© 1998 by CRC Press LLC
Lewis Publishers is an imprint of CRC Press LLC

No claim to original U.S. Government works
International Standard Book Number 1-56670-295-X
Library of Congress Card Number 97-27474
Printed in the United States of America 1 2 3 4 5 6 7 8 9 0
Printed on acid-free paper

Dedication

This book is dedicated to my lovely wife Dawn, and my three wonderful children, Elliott, Zachary, and Nicole. This is what I was doing all of those nights on the computer.

PREFACE

How do we move out of the reactive mode of maintaining environmental compliance? How do we integrate Total Quality Management (TQM) principles into a technical area such as environmental compliance? How do we make pollution prevention and environmental ethics part of our organizational culture? How can we expend less resources on compliance while still reducing our overall impact on the environment?

I've sought answers to these questions for many years, but found no single publication which told me how I can move beyond compliance. I did find numerous publications which provided various parts of this puzzle, but none hinted at integrating the pieces. With the emergence of the ISO 14000 standards I was confident I had found a solution, but again the practical application seemed elusive.

After much research, I decided to develop a handbook to help guide organizations to reduce their impact on the environment while also reducing operating costs associated with maintaining environmental compliance. I realized that at the core of this discussion are the principles of pollution prevention and continuous improvement. This is a "hands-on", "how-to" book focused on applying the philosophy discussed in many other texts and publications currently on the market today. Within this text you'll find suggested methods, tools, and techniques for integrating pollution prevention, ISO 14000, and environmental compliance, so that you can reduce your burdens and have cost-effective environmental programs.

You will notice that I have kept this text as informal as possible; in fact it is best used as a handbook to help you along your environmental journey. As I wrote this document I envisioned that I was making a presentation to you, the reader, just as if you were sitting in a classroom or a seminar room. The text follows a very methodical, step-by-step, approach to integrating ISO 14000 and pollution prevention. If you view this integration as a journey you realize that each step is simply a milestone of your journey. Focusing on and obtaining these milestones will keep you and your organization on track to meet the goals you set.

The first two chapters are somewhat basic — I wanted to ensure that all readers have a general understanding of the magnitude of the environmental compliance issues facing organizations, as well as a general understanding of the ISO 14000 standards. But this is not a textbook on compliance or the standards — many good books and other references are available on this subject. In fact, if you need more information, numerous references are provided in Chapter 12.

As the book develops you will learn how to properly conduct and apply the three basic steps of implementing and maintaining an effective pollution prevention program, the vulnerability assessment, baseline assessment, and the opportunity assessment. Additionally you will learn about many tools you can use to screen your processes down to only a vital few and realize the key to success is maintaining focus on only a couple of opportunities at a time. You will learn how to integrate these steps with ISO 14000 and to how to use performance measurements and environmental audits to foster a waste-reduction mindset within your organization. Finally, I have provided numerous detailed checklists and plenty of reference information, which should be helpful as you journey to move beyond environmental compliance.

I wish you luck on your journey and hope you find this information very helpful.

List of Figures

Contents

Chapter 3

Chapter 7

Chapter 8

Chapter 11

1 An Idea Whose Time Has Come

We all agree with the need to protect the environment. As we have industrialized to increase our standard of living, our environment has suffered. The resultant environmental regulations have been successful in protecting human health, restoring the natural environment, and protecting our natural resources, but this success has come with a cost. This cost is an increase in operating expenses necessary to maintain compliance with the numerous complex and sometimes conflicting environmental regulations, and we have found that achieving and maintaining compliance with the numerous environmental regulations is just plain hard to do!

As the number of federal, state, and local laws and regulations continues to increase, so will the cost of doing business. Regardless of the type of business, whether it's manufacturing or a service organization, we all pay the cost for environmental safety and health issues. This cost is hidden in a number of ways; ever-growing environmental staffs, increasing dependence on consultants, increasing training requirements, plus the increasing costs of protecting individuals from the harsh chemicals we use in our day-to-day business.

Although the laws, regulations, and standards are absolutely essential for our long-term health and the protection of the natural environment, the financial cost to comply is huge. This ever-increasing cost of environmental compliance is passed on to consumers, cuts into profits, and reduces an organization's competitive edge. At the very least, the cost of compliance cuts into operating budgets, and directs attention away from the primary business of the organization. These expenses have heightened interest from all types of organizations to move beyond compliance in a proactive manner to reduce their impact on the environment while reducing the operating expenses of maintaining compliance.

Although this concept of managing to minimize the impact on the environment is relatively new to many organizations, some have been using these concepts for years. These organizations have realized significant improvements in operational performance and environmental protection. Today more and more organizations include the environment in determining what constitutes a successful operation along with the typical management considerations. These organizations have learned that there is a way to have both environmentally safe business systems and low operating costs.

By developing effective environmental management systems, such as those described under ISO 14000, and by applying pollution prevention methodologies, your organization can also drastically reduce your compliance burden. By following the methodology laid out in this text you'll be able to reduce waste generation to the point where the hassles of maintaining compliance are eliminated, or at the very least significantly reduced.

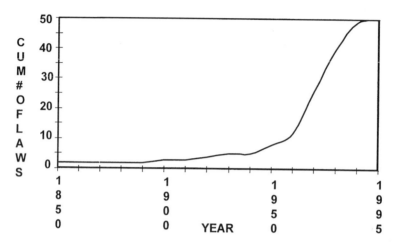

FIGURE 1.1   The progression of environmental laws.

Before we discuss these principles and methods, let's first look at the history of environmental compliance so you will understand that managing with the environment in mind is simply the next logical step in the environmental journey.

A BRIEF HISTORY OF ENVIRONMENTAL REGULATION[1]

As the public became concerned with depleting natural resources, increasing hazards to human health, and protecting the environment, Congress responded with the passage of numerous laws. As the number and complexity of laws increased, so did the difficulty and expense of maintaining compliance with the numerous federal, state, and local laws, regulations and ordinances. As an example, it's estimated that complying with the Clean Air Act and its amendments costs U.S. industries over $10 billion annually.[2]

Figure 1.1 clearly illustrates the progression of environmental laws through the years and the growing wave of federal environmental regulation in the last 30 years. In fact, 72% of the laws depicted in this figure were passed after 1965. The laws now represent over 10,000 pages of federal environmental regulations.

You may be wondering why the sudden awareness of environmental concerns 30 years ago? Many point to the cultural revolution of the 1960s and one particular book, Rachel Carson's *Silent Spring*. *Silent Spring* tells the story of commonly used pesticides, such as dichloro-diphenyl-trichloro-ethane (better known as DDT) poisoning the water and decimating animal populations as the chemical bio-magnifies up the food chain. *Silent Spring* awakened a nation to a new environmental ethic and heightened public awareness to what industry and their fellow men were dumping into the environment around them. With the proliferation of environmental laws, and a more environmentally aware public, the federal agencies also increased their awareness. What was once common practice, and legal, now seems unthinkable. For instance, a 1946 Army Air Corps manual[3] illustrated how firefighters trained for aircraft fires. They dug a large pit and then filled it with water, then simply poured aviation fuel on top and set it ablaze. The resulting inferno, of course, put an acrid black cloud into the air, and the unburned fuel sunk into the ground. Today's environmentally compatible firefighting "pits" are fully lined with membranes and use liquefied propane gas (LPG) as the fuel. LPG produces no cloud of smoke and comparatively no pollution. Water used to fight fires is captured, recycled, and not allowed to enter the ground. In other instances, personnel once poured waste solvents into unlined evaporation pits where they were thought to be harmlessly evaporating into the atmosphere. The solvents were evaporating, but they were also sinking into the ground and contaminating groundwater. Today, waste solvents are properly handled as hazardous waste, with rigorous procedures to prevent any from getting into the ground.

THE BIG FIVE

Most organizations are affected by five laws passed and amended in the last 30 years: the Clean Air Act (CAA), the Resource Conservation and Recovery Act (RCRA), the Safe Drinking Water Act (SDWA), the Toxic Substance Control Act (TSCA), and the Clean Water Act (CWA). Compliance with these laws require technical experts on your staff, extensive training, and numerous permits and plans. Let's briefly review each of these laws so we have a better understanding of their impacts on your organization.

CLEAN AIR ACT

The Clean Air Act (CAA) was passed in 1963 and had been amended 16 times by 1990. The CAA established regions within the United States where the air does or does not meet national ambient air quality standards (NAAQS). A region either attains NAAQS or is in "non-attainment." For instance, Los Angeles, California, currently sits in a non-attainment region. For regions of non-attainment, the states and local air quality management districts must promulgate a state implementation plan (SIP) which implements the CAA. Organizations within non-attainment regions must abide by and conform to SIPs. Any substantial organizational growth must ensure it does not adversely affect the ambient air after expansion. Compliance with this law is extremely labor intensive. Numerous permits, studies, and process analysis are required to meet the regulatory requirements driven by this series of laws.

RESOURCE CONSERVATION AND RECOVERY ACT

The Resource Conservation and Recovery Act (RCRA) focused on the generation, storage, treatment, transportation, and disposal of hazardous waste. The Environmental Protection Agency (EPA) defined hazardous wastes as those wastes that either corrode, or react violently with their surroundings, or are toxic to living organisms, or are ignitable, or are listed as hazardous by government regulations. Congress passed RCRA as an amendment to the Solid Waste Disposal Act in 1976. RCRA, as amended in 1984 by the Hazardous and Solid Waste Amendments (HSWA), now stands as a tremendously complicated and onerous law that affects virtually all American municipal, commercial, and industrial activities.

To further emphasize the impact of RCRA, in 1992, Congress passed the Federal Facilities Compliance Act (FFCA). FFCA basically stripped all federal agencies of their sovereign immunity from the full impact of federal and state solid and hazardous waste laws. It made them financially vulnerable to environmental enforcement agencies such as EPA or any state environmental agency. In other words, federal agencies could be directly fined for failing to comply with hazardous waste laws, and the fine would come out of their operating budgets.

Compliance with RCRA is also labor intensive. It requires a great deal of training, permitting, and continuous internal oversight. As with the Clean Air Act, most hazardous wastes are generated through processes which use hazardous materials. In general terms, eliminating the use of a single hazardous material can significantly reduce your responsibilities under RCRA.

SAFE DRINKING WATER ACT

The Safe Drinking Water Act (SDWA), passed in 1974 and, as alluded to earlier, placed stringent and expensive requirements upon the quality of the water we drink. Many organizations purchase their water from public utilities. These utilities bear responsibility for the quality of the water; however, many organizations ensure the water meets SDWA standards. In addition, meeting the SDWA requirements drove up the cost of water bought from local communities and placed further monitoring and testing restrictions on already over-tasked staffs. Although there are many regulatory requirements under the SDWA, most industrial organizations are not greatly affected by them. They cannot reduce their compliance-related workload without significant changes to their manufacturing processes.

TOXIC SUBSTANCES CONTROL ACT

The Toxic Substances Control Act (TSCA), passed in 1976, set standards for the manufacture and use of certain hazardous substances. Most important to most organizations were asbestos and polychlorinated biphenyls or PCBs. Asbestos, once hailed as a miracle building material because of its fireproof characteristics, became an ugly word to many facility maintainers. Scientists linked asbestos to respiratory ailments and other diseases including lung cancer. The electrical utility industry used PCBs, hailed for its fireproof characteristics, for years as the dielectric fluid in electrical transformers and capacitors. PCBs, also linked to cancer, appear in the tissues of fish and birds.

Although this law can cause a great deal of work for your organization, the work can be eliminated with the removal of PCB-containing electrical equipment and the removal of all asbestos-containing building materials within your facilities. Although this is no small task, it is achievable over time with a focused effort on the part of senior management.

CLEAN WATER ACT

The Clean Water Act (CWA) deals primarily with the quality of wastewater returned to the environment. After its passage in 1977, millions of dollars flowed to municipalities to upgrade wastewater treatment plants to meet new discharge standards. The CWA also covers the use and protection of wetlands. Areas that meet the federal definition of "wetlands" exist on many facilities and must be properly managed. In some cases, this limits the ability to carry out planned expansion. There are many regulatory requirements under the CWA. Most industrial organizations can immediately reduce their compliance related workload by analyzing the materials used in their processes which ultimately end up in their waste streams.

From our discussion above, you can minimize the effect of three laws: the CAA, RCRA and to some extent, the CWA, by using less toxic materials in your manufacturing processes. Throughout the history of environmental regulation most organizations have made every attempt to comply with all laws as they existed at the time. Good-faith efforts, made by those in charge at all levels, demonstrated them to be good stewards of the environment. As mentioned earlier with the fire protection training and the solvent disposal examples, we all contribute to the problem of pollution; however, at the time that these activities were ongoing, they were "legal."

Due to changes in environmental regulations and an expanding awareness for environmental matters, environmental compliance efforts also expanded to a level of sophistication that we know today: compliance-based environmental management systems.

THE COMPLIANCE-BASED EMS

An Environmental Management System (EMS) is composed of that aspect of an organization's overall management function that determines and implements the organization's environmental policy. Such a system can be described or characterized in terms of the organizational structure, assigned responsibilities, practices, procedures, policy, goals, and objectives. Although an EMS can be designed to achieve a range of goals and objectives, most systems are designed, or at the very least operated, to meet the requirements of and respond to laws and regulations enacted by federal, state, and local authority, i.e., they are compliance-based.

With a compliance-based EMS the organization environmental program has been structured to achieve compliance with the environmental regulatory structure. The character of the regulatory structure has been shaped by the basic assumption that the surest and best way to remediate, minimize, or eliminate the negative environmental effects of human behavior and production is by enforcing a body of laws and regulations and by applying coercive measures to assure compliance. Historically, the United States responds to new environmental problems with a new law, which usually requires a new set of regulations and programs such as those discussed earlier. Consequently, the U.S. environmental management system is a regulation-driven, compliance-based, programmatic approach to environmental protection. The system is also characterized by its command-and-control structure of complex laws, oversight, and enforcement authorities and penalties.

Process	Characteristic
Driver	Laws and Regulations
Perceived Responsibility	Compliance
Policy	Compliance
Approach	Programmatic (air quality, water quality, hazardous waste management)
Management Structure	Command and Control (inspection, laws and penalties)
Goals	Full Compliance
Measurements	Environmental system results (i.e. chemical releases, permit violations). Operating systems results (i.e. waste by-product per unit of production).
Measurement Tool	Environmental Compliance Audits

FIGURE 1.2   The compliance-based EMS model.

The compliance-based EMS is not effective for assuring continuous improvements in environmental performance and for demonstrating environmental achievements. Although the compliance-based environmental management system evolved in response to the immediate need to be compliant with regulatory requirements, it is not a complete system. It lacks the tools and structure to improve compliance performance, minimize compliance costs, decrease regulatory oversight, reduce or eliminate penalties, and to demonstrate environmental achievements. While compliance-based environmental management systems have resulted in substantial improvements in environmental quality and human health and safety, those systems are not developing into more progressive EMS standards. Figure 1.2 summarizes the compliance-based EMS's as reactive, regulatory-driven systems designed to deal with programmatic and operational regulatory requirements. The time has come to move from compliance-based management systems into the next step in the environmental journey; a mangement system based on prevention, and continuous improvement as shown in Figure 1.3.

Process	Characteristics
Driver	Environmental and economic efficiency
Perceived Responsibility	Environmental Protection
Policy	Compliance, prevention continuous improvement, and leadership
Approach	Systematic and strategic
Management Structure	Systematic, integrated, quality-based systems
Goals	Full compliance, pollution prevention, health and safety, and resource conservation
Measurements	Environmental system results (i.e. chemical releases, permit violations). Operating systems results (i.e. waste by-product per unit of production). Management systems results (i.e. organizational structure, staff, training, reporting, planning and risk management)
Measurement Tool	Environmental Compliance Audits, EMS Audits, internal process measures.

FIGURE 1.3 The continuous improvement based EMS model.

UNDERSTANDING THE TOTAL COST OF ENVIRONMENTAL COMPLIANCE

As you have already seen, the cost of environmental compliance nationally is huge. As an example, let's take a closer look at the total cost of using a hazardous material with respect to various environmental compliance regulations. This example shows that the actual disposal of the waste is simply the "tip of the iceberg" (Figure 1.4). The total costs are much greater, but are typically hidden in the sea of operating expenses. These total costs include the costs of using a hazardous or toxic material and the costs of protecting the user, in addition to the costs associated with disposal activities.

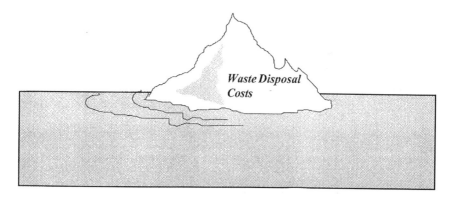

FIGURE 1.4   Hazardous waste disposal costs: the tip of the iceberg.

Let us further evaluate these costs. The costs of using hazardous or toxic materials include:

- The initial purchasing price
- Expense to ship in a "DOT" approved manner to include packaging and shipment
- Expense of proper warehousing or storage locations
- The cost of training and personal protective equipment (PPE) for the warehousing personnel
- Spill prevention and response kits within the warehouse
- OSHA compliance requirements to include: EPCRA/SARA Title III reporting requirements, health and safety training personnel and inspection, shipping and storage,
- Personal Protective Equipment (PPE) to use the hazardous material
- Facility upgrades such as proper ventilation
- Personnel training
- Spill response teams to include training and equipage
- Environmental permits
- Environmental compliance experts on your staff

The costs of disposal of the waste or unused material include:

- Hazardous waste disposal fees
- Shipping fees
- Environmental compliance experts to complete manifests
- Storage of manifests
- Regulatory reporting
- Environmental compliance experts on staff
- Life-long liability chain

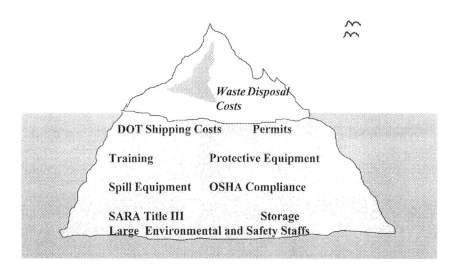

FIGURE 1.5   Below the surface: the hidden cost of hazardous material.

When you evaluate the total costs of using hazardous materials (Figure 1.5), rather than just evaluating the cost of waste disposal, you begin to understand the enormous savings which can be realized by eliminating or reducing the use of a single hazardous material. Can you imagine the savings if you eliminated all of the hazardous material throughout your entire organization? This basic question has heightened awareness in both private and public sectors. Company and government executives have begun to look for a tool to help ensure compliance and avoid costly fines or adverse publicity associated with noncompliance. These tools include pollution prevention and more holistic environmental management systems.

AVOIDING THE HASSLES OF ENVIRONMENTAL REGULATIONS

The compliance-based systems have been necessary and effective in protecting human health and the environment, but it's time to move to a more complete environmental management system. Today most environmental management experts believe that compliance-based command and control systems have reached the point of diminishing returns.[4] Compliance-based systems are not flexible, consistent, or efficient enough to achieve goals that go beyond compliance and lead to lower environmental protection costs. Many changes, both regulatory and nonregulatory, are creating the need for a more flexible, consistent, and efficient EMS. These factors include:

- Shifting public concern and awareness.
- Perception shifted from compliance to stewardship.
- Environmental goals aimed at pollution prevention.

- Environmental management principles, when based on stewardship, require proactive rather than reactive management approaches.
- Budget and personnel staff reductions require better performance at lower costs.
- New international trade agreements require consistent approaches between countries.

Any management system which goes beyond compliance can also result in greater regulatory flexibility and public goodwill at a time of increasing environmental awareness from the general public. Numerous models for environmental management have emerged and are widely supported by environmental advocates, trade, and professional organizations. These models implement a systems approach based on continuous improvement and environmental stewardship processes, an approach that goes beyond compliance and integrates pollution prevention and the principles of Total Quality Management (TQM). These Environmental Management Systems (EMS), such as those developed under ISO 14000 and other environmental codes, are typically referred to as Total Quality Environmental Management (TQEM) initiatives,[5] and when properly applied can significantly reduce the cost of maintaining compliance with the numerous environmental laws discussed earlier, through reducing your organization's impact on the environment.

Successful Pollution Prevention (P2) programs require an organization to refocus its environmental policies from a reactive compliance-based policy to a more proactive "prevention"-based policy. This is consistent with the concept developed under the ISO 14000 Standards. Combining these concepts with the powerful management tools and techniques taught to us by many TQM professionals, we can overcome the burdensome environmental compliance issues and refocus on the primary business line of the organization. By following the principles, examples, and descriptions I have presented within this text you will greatly reduce the operational costs of maintaining compliance with environmental and safety regulations. Additionally, the checklists and examples provided throughout the text will provide additional tools for your environmental safety and health professionals.

WHAT IS POLLUTION PREVENTION?

The EPA describes Pollution Prevention, or P2 as I will refer to it throughout this text, as "the use of materials, processes, or practices that reduce or eliminate the creation of pollutants or wastes at the source.[6]" It includes practices that reduce the use of hazardous materials, energy, water, or other natural resources and practices that protect natural resources through conservation or more efficient use. The primary focus is on reduction or elimination at the source, followed by recycling, treatment, and finally disposal as a last resort. This is typically referred to as the P2 hierarchy, or the pillars of prevention, as shown in Figure 1.6.

P2 efforts usually focus on reducing the use of hazardous materials, thus reducing the amount of hazardous waste and hazardous air pollutants generated by a facility. But P2 efforts are not limited to just hazardous materials. There are also great

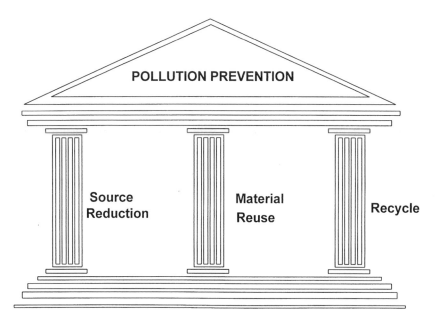

FIGURE 1.6   Pillars of pollution prevention.

opportunities to reduce solid waste, air emissions, finding better business practices, and reducing energy and water consumption.

P2 efforts coincide directly with organizational efforts to reduce operating costs and the "hassles" created with environmental compliance. A company with an effective continuous P2 program will greatly reduce its operating costs, thus potentially creating a market advantage over its competition. The costs savings will be realized through reductions in operating costs and reductions in environmental liabilities.

Operating costs include material costs, production costs, disposal costs, and energy and water costs. Material costs are reduced when more efficient packing and production methods are used, thereby producing less waste. Waste management and disposal costs will decrease as P2 initiatives are implemented. These cost savings include:

- Reduced manpower and equipment requirements for on-site pollution control and treatment
- Less waste-storage space
- Less pre-treatment and packaging prior to disposal
- Smaller waste quantities treated
- Lower waste-production taxes
- Reduced paperwork and record keeping

Production costs can also be reduced through the identification of more efficient processes. Production scheduling, material handling, inventory control, and equipment maintenance are some of the areas where process analysis can identify the potential for greater efficiencies. Other operating costs which can be reduced with pollution prevention include a reduction in energy and water expenses.

Liability will be reduced as you reduce the amount and toxicity of your waste. Without the waste you will not be subject to as many environmental laws and regulations, which include criminal and civil liabilities. Also, your workers' risk will be reduced along with worker compensation costs.

When you view the pillars of prevention, you see that source reduction receives the highest priority. Source reduction is defined[7] as any practice that reduces the amount of any hazardous substance, pollutant, or contaminant entering any waste stream or otherwise released into the environment prior to recycling, treatment, or disposal; and reduces the hazards to public health and the environment associated with the release of such substances, pollutants, or contaminants. Source reduction includes equipment or technology modifications, process or procedure changes, reformulation or redesign of products, substitution of raw materials, and improvements in housekeeping, maintenance, training, or inventory control.

For organizations just starting on the P2 journey, most prevention projects pay for themselves within a year, with more complex projects taking under 5 years to achieve a return on the investment. In fact, many P2 initiatives are so simple that there are minimal overhead costs; simply process changes.[8] Once these easy projects are completed more comprehensive projects may be required. By following the methodology discussed within this text you'll be able to systematically achieve effective P2 implementation on both simple and complex projects.

ENVIRONMENTAL MANAGEMENT AND TOTAL QUALITY

Quality management and P2 go hand-in-hand and are similar in many ways. For starters, both principles follow the continuous improvement mind-set. TQM is the application of continuous improvement of processes to improve performance, reduce cost, and increase efficiency. P2 is a continuous review and study of all waste-generating activities to minimize waste generation. Both principles also involve people at the operational level to identify and implement solutions to reduce long-term operational costs and to drive *"work and time"* out of our processes. P2 borrows from TQM the systematic analysis of business processes by empowered cross-functional teams. Waste identification and reduction has proven to be most successful when line employees are teamed with engineering and management personnel to identify and implement innovative waste- and cost-reducing concepts. This may include purchasing new equipment and using environmentally safer chemicals that not only drive down environmental costs, but also lessen the labor used in the process. A third way TQM and P2 are similar is through increasing the performance of internal process by reducing waste within the process itself. You constantly improve the "quality" of the product or service your company provides. This, of course, will decrease the cost to manufacture, decrease the defect rate, and increase market share. This large "Q" refers not only to increasing the overall quality of the product, but also the cost to produce the product. Yet another way P2 is similar to quality

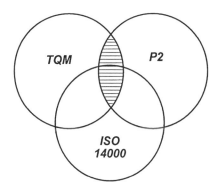

FIGURE 1.7   The Venn diagram.

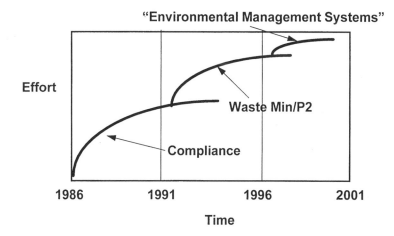

FIGURE 1.8   The environmental journey continues.

management is through the application of setting goals and objectives and using a measurement system to track your improvements or reductions (Figure 1.7).

So how does ISO 14000 fit into this? ISO 14000 is one of many environmental management standards requiring the development of an Environmental Management System (EMS) which focuses an organization's leadership on viewing their actions as to how they impact the environment. It requires senior management attention in policy making, goal setting, and measuring success in achieving those goals. Using a proactive approach to the waste-generating system established by your organization allows you to place your attention beyond compliance and onto prevention. As you can see, this is similar to the discussion we had above regarding P2 and Quality Management. In fact, I like to view the EMS as simply the next step in the environmental journey as shown in Figure 1.8. Armed with this background let's now review ISO 14000.

ISO 14000 HISTORY[2]

The International Organization for Standardization (ISO) is a worldwide federation that promotes the development of international manufacturing, trade, and communication standards. ISO is best known by most of us for the numbering system on the film speed we use in cameras and for the 9000 series quality-management standards. The ISO representative in the United States is the American National Standards Institute (ANSI).

In 1991, ISO established the Strategic Advisory Group on the Environment (SAGE) to make recommendations on the international environmental management standards. In 1993, SAGE recommended that a Technical Committee (TC) be formed to develop the EMS standard, and TC 207 was formed to meet this need. This is typical, as international standards are normally developed by technical committees. Member nations interested in a subject are represented on committee working groups. Committee recommendations are reviewed by government, industry, and other public and private sectors before a standard is formalized. Implementation of ISO standards is voluntary, although most international governments require compliance for trade within their countries.

Prior to developing the 14000 Standards, the ISO developed the 9000 series Quality Management Standards. The International Quality Standards (ISO 9000) were developed in 1987 to meet a need for common standards that would allow potential users to effectively evaluate products and services. It is used by 60 countries, including the United States, Canada, and Mexico. The ISO 9000 quality standards provide a framework for incorporating all the elements required to select, implement, manage, and monitor quality across a multinational environment. These elements include traceability, documentation, audits, training, and management review. Some reasons organizations seek ISO 9000 registration include:

- Help compete in international markets
- Show commitment to quality
- Gain strategic advantage over non-certified competitors
- Meet customer requirements

The ISO 14000 movement solidified in the United States on January 18, 1996. The ANSI Technical Advisory Group 207 met in Philadelphia to announce that five standards would be initially adopted and published as U.S. National Standards after final approval by the ISO committees. Federal and state agencies are now reviewing ISO 14000 and the driving forces which are causing the private sector to adopt it as well.

ISO 14000 OVERVIEW

ISO 14000 is a voluntary standard based on the fact that people and companies want to do the right thing for the environment while improving operational efficiencies. As I discussed earlier in this chapter, many organizations have determined that by systematically integrating environmental concerns into overall management activities

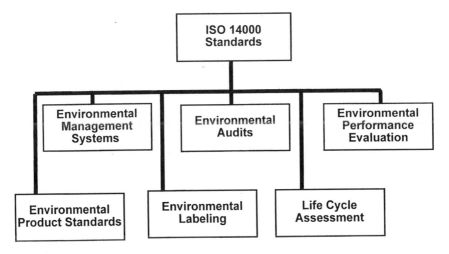

FIGURE 1.9   The six ISO 14000 standards.

such as strategic planning and policy making, they can achieve significant reductions in operating costs while exceeding compliance requirements.

The ISO 14000 Standards are actually a total of six standards that are described in detail in Chapter 2 and shown below. The standards include (Figure 1.9):

- Environmental Management Systems
- Environmental Performance Evaluations
- Environmental Audits
- Life Cycle Assessments
- Environmental Labeling
- Product Standards

In general terms, the standards require that an organization have a system in place for its subordinate units, such as plants, divisions, or other type of business units, to meet (or exceed) previously established environmental goals without oversight from the corporate environmental staff. To do this, the first step is for senior management to develop a formal environmental policy that includes:

- Protecting the environment as one of the key organizational priorities. It should also include clear assignment of responsibilities and accountabilities to all employees.
- Maintaining compliance with all applicable environmental laws and regulations relative to your activities, services, and products.
- Ongoing communications on environmental commitment and performance with all stakeholders.
- Strategic planning efforts that develop environmental performance objectives and targets and which are implemented though a systematic, continuous improvement management approach.

- Regular performance updates, including audits and management reviews to achieve continuous improvements.
- Fully integrating environmental activities into other functions and processes.

You may be asking why you should expend the resources to understand ISO 14000? You should, primarily because it is going to affect the way business is conducted globally in the near future. The standard will first affect U.S. interests operating abroad, then will impact companies within the U.S. By getting on board now, you can proactively work the issues now, rather than react to them later. Some specific reasons include:

- ISO 14000 Standards and certification may become a requirement for doing business in Europe.
- ISO 14000 provides an opportunity to improve your environmental management systems to go beyond compliance and prevention in an effort to integrate operational and environmental requirements. In short, it's the next step in the environmental journey.
- ISO 14000 will greatly affect the U.S. environmental market just as ISO 9000 affected the manufacturing sectors.
- Equally important are the significant performance improvements that can be gained by having an effective EMS in place. An effective EMS can help you focus on prevention and
   Reduce the costs of maintaining compliance,
   Reduce or minimize legal liabilities,
   Reduce operating costs and financial liabilities,
   Enhance stakeholder image, and,
   Obtain access to markets.

The ISO 14000 model when integrated with an effective pollution prevention effort should allow numerous organizations to realize a competitive advantage of having superior environmental performance. This will occur because of reduced operating costs through more enviromentally sound business practices. ISO 14000 provides the basic foundation to move beyond the hassles of compliance to improve the environmental performance of your activities.

CLOSING THOUGHTS

Integrating environmental concerns into day-to-day organizational management is the next logical step in a journey which began with environmental compliance, then grew into prevention. Within the United States, we are seeing an increasing awareness in environmental codes and standards, especially the ISO 14000 Standards, as being the tools to help move beyond compliance while also reducing impacts to the environment. This is a very positive step, but I'm concerned that many organizations will focus their attention on developing very detailed documentation packages that meet ISO 14000 requirements and lose focus on their prevention activities. Typically, P2 activities have focused on waste-minimization activities. These are effective at reducing operating costs and enhancing cost-avoidance opportunities. But the real pay-back occurs when you begin analyzing your processes as "systems." By analyzing the entire waste-generating system, you can better understand the actual causes of the inefficiencies and attack those root causes. As the next few chapters unfold, I'll show you how to use the ISO 14000 framework to fully integrate environment management and pollution prevention so you can reduce your impact on the environment. Always keep in mind that the objective of all your activity is to eliminate inefficiencies that generate waste during production or product use. Reducing the inefficiencies will reduce your operating costs, reduce the hassles, and hopefully will also create a competitive advantage. Before you consider the EMS journey to achieve ISO 14000 certification, you should have a very mature and stable compliance program and a fully functioning prevention program. Then, once these are in place, you can follow the sequential steps laid out within this text to help you move beyond compliance to a more mature EMS while reducing the operating costs and hassles of maintaining compliance.

REFERENCES AND NOTES

1. Fant, R., and others, *The Air Force's Environmental Compliance Assessment and Management Program: A Comprehensive Review and a Direction for the Future,* USAF internal document, (available from the Air University Library, Maxwell AFB, Alabama), April 1996. Much of the discussion on the history of environmental compliance was provided by my friend and colleague, Bob Fant. His research in regulatory history and on environmental audits provided a refreshing and insightful review of the environmental journey up through 1995.
2. Dr. Joe Fernando and Tom Welch, *ISO 14000, Emerging Environmental Management Standards,* p. 2, 9, Brooks AFB, San Antonio TX, February 1996. This paper was developed to provide an overview of the ISO 14000 Standards to federal agencies.
3. 1946 Army Air Corps Field Manual, *Firefighters Training Manual,* p. 4-3.
4. Strategies for Improving Environmental Management in the DOD, Office of the Inspector General, Department of Defense, August 1996.
5. Conversations with Dr. Charles Giamona, Executive Director of REM, a Research, Extension and Management Corporation associated with Texas A&M. While developing this text, Dr. Giammona provided great insight into the application of Environmental Management Systems. Also, the term "TQEM" is typically recognized as first used by GEMI, as discussed in Chapter 3.
6. EPA Document Number 300 B-94-007, *Pollution Prevention for the Federal Facilities,* p. 1.
7. EPA Document Number 300 B-94-007, *Pollution Prevention for the Federal Government,* p. 4.
8. Dougherty, J. and Welch, T. Implementing an Effective Pollution Prevention Program, U.S. Air Force, USAF Environmental Symposium Proceedings, Feb. 1993.

2 An Overview of ISO 14000 and Other Environmental Standards

DEVELOPMENT OF STANDARDS AND CODES

ISO 14000 is just one of many voluntary environmental standards and codes that have emerged since the mid 1980s. These standards have grown out of the increased public concern with environmental issues and corporate concerns with the ever-increasing number of environmental regulations and environmental compliance costs.

In England, the British Standards Institute developed BS 7750, an environmental management system (EMS), which is accepted by many companies both in England and elsewhere. The European Union adopted the Eco-Management and Audit Scheme (EMAS), which specifies the EMS for companies doing business in the European Union. Other national EMS standard efforts are being developed in many countries, including Canada, France, Spain, and Ireland. Other standards or codes include:

- The ISO 14000 international standards
- The Business Charter for Sustainable Development developed by the International Chamber of Commerce
- Responsible Care Program developed by the Chemical Manufacturers Association (CMA)
- Coalition for Environmentally Responsible Economies (CERES) principles
- The Global Environmental Management Initiative (GEMI)
- The ANSI/ASQC (E-4) "Systems and Guidelines for Quality Systems and Environmental Data Collection and Environmental Technology Programs"
- The EPA's Code of Environmental Management Principles

While each of these standards has strengths and particular focuses, they also have many elements in common. Each of these standards requires an organization:

- To establish an environmental management system (EMS).
- To audit their EMS to determine if they are achieving the goals they set for themselves.
- To evaluate their product's impact on the environment based on life cycle management.
- To involve outside groups in their environmental efforts. These groups can include the local community, but typically focus around the customers and suppliers to the organization.

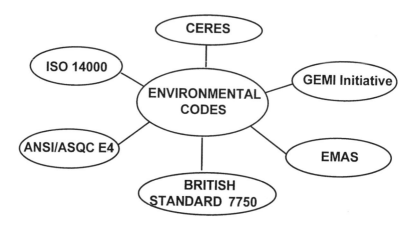

FIGURE 2.1   Some of the numerous environmental codes and standards.

It important to note that none of these voluntary standards have any type of performance-related "compliance" standards that the organization is required to meet. Instead they are designed to help organizations develop an ethic which makes them consider their impact on the environment.

So if organizations are not driven by regulatory pressure or by performance-related issues, why would they decide to add that seeming additional burden onto themselves? These voluntary codes can help to:

- Reduce the cost of doing business.
- Distinguish yourself as an environmental leader.
- Create consistency across various plants and various regulatory agencies.
- Increase positive public relations.

I mentioned above how each of the standards tends to center on four basic concepts. This, I believe, is an indicator of how environmental management has grown, or continues to grow, beyond the requirements driven by regulatory agencies to meet the expectations of the public and of customers. Each of these codes and standards really focuses on four basic concepts: EMS, life cycle management, commitment to continuous improvement, and interaction with all stakeholders. Let's briefly look at each of these:

ENVIRONMENTAL MANAGEMENT SYSTEMS

An Environmental Management System (EMS) is that aspect of an organization's overall management function that determines and implements the organization's environmental policy. Such a system can be described or characterized in terms of the organizational structure, assigned responsibilities, practices, procedures, policy, goals, and objectives. Essentially, an EMS is how your organizational managers go about identifying, addressing, and correcting environmental problems. Typically this includes such things as assessing environmental concerns, establishing targets and goals, developing policy statements, training workers, auditing how you are doing against the policies, goals, and targets; and sometimes it includes having a third party reviewing your progress.

As I discussed in Chapter 1, the concept behind an EMS is for senior management to specify those core environmental requirements for their organizational management systems and not to state performance criteria such as compliance criteria. Numerous proactive organizations have understood this principle and now have a competitive advantage because effective environmental management systems are in place, thus reducing operating expenses. One of the cornerstone principles of an effective EMS has been on, and continues to be on, pollution prevention and waste minimization activities. This is because these activities reduce operating costs and enhance cost avoidance opportunities. Equally important are the other significant performance improvements that can be gained by having an effective EMS in place. These include:[1]

- Reducing the costs of maintaining compliance
- Allowing you to focus on prevention
- Helping reduce or minimize legal liabilities
- Reducing operating costs and financial liabilities
- Enhancing stakeholder image
- Gaining competitive advantage
- Obtaining access to markets

In a perfect system, we would generate no waste and have no adverse impact on human health or the environment. Our overall goal is to eliminate the significant wastes and impacts. Obviously, as we achieve this goal and reduce our environmental costs, this results in lower costs of products or services relative to our competitors, therefore allowing gains in market share.

Life Cycle Management

Life cycle management typically refers to how an organization reduces its impact on the environment through all stages of a product's life. This is essentially a "cradle-to-grave" approach to environmental impacts.[2] The typical environmental compliance issues and certainly most pollution prevention activities focus on the manufacturing process itself, but life cycle assessment complements these by evaluating the impacts on the environment for the entire life of the product. This begins with the supplies and raw materials going into the process, followed by manufacture, transportation, operation and maintenance of the product, associated waste generated by the product, and finally disposal or reuse. This is the "life-cycle" perspective of products and processes, where you consider the impact to the environment not only in terms of waste generated during production, but also during product use and disposal to ensure that you really analyze your overall operating efficiencies and reduce your long-term liabilities.

COMMITMENT TO PROTECTION AND CONTINUOUS IMPROVEMENT

Most codes require senior management to develop a policy and define its responsibility to protect the natural environment. As with many TQM-based programs, the concept of continuous improvement is not limited to simply manufacturing products. The concept is applied to creating ongoing reductions in waste-generation and of course links directly to the goals and objectives developed in the EMS. This also may include such things as reducing energy use and more efficient use of natural or other resources.

INVOLVING ALL STAKEHOLDERS

Many of the codes recommend identifying and involving all of the process stakeholders, including the public and local community, with the environmental program. This encourages buy-in from employees and the community in all the environment-related decision making. Stakeholders need to include the employees and process owners who have the most to gain from implementing successful solutions.

VARIOUS CODES

While it's beyond the scope of this text to discuss the details of each code and standard, I'd like to briefly review some of them. Then I'll focus on the ISO 14000 standard to develop the linkage to pollution prevention throughout the text. It's important to note that virtually any of the standards can be applied to the models I develop further in this text.

TOTAL QUALITY ENVIRONMENTAL MANAGEMENT (TQEM) FROM THE GLOBAL ENVIRONMENTAL MANAGEMENT INSTITUTE (GEMI)

GEMI is a group of about 27 leading companies whose members collaborate in efforts to promote a worldwide business ethic for environmental management and sustainable development, to improve the environmental performance of business through example and leadership, and to enhance the dialogue between business and its public. GEMI is generally credited as being the first organization to marry environmental management and the total-quality-management philosophy: plan-do-check-act. The GEMI process is identified as total quality environmental management (TQEM), a systematic approach to continuously improve environmental performance.

ENVIRONMENTAL SELF-ASSESSMENT

The environmental self-assessment program is based on principles published by the International Chamber of Commerce's Charter for Sustainable Development. The environmental self-assessment program serves as a tool for improving an organization's environmental management system and environmental performance. The program is designed to measure and analyze corporate environmental management performance over time, so business can pinpoint ways to increase the quality of environmental policy, planning, implementation, and monitoring.

ASTM EMS STANDARDS MODEL

This model was designed for the U.S. market, based on two standards documents. The first document, "Practice for Environmental Compliance Audits," gives a description of accepted practices, procedures, and policies associated with environmental compliance audits — assessments. This document, based on EPA and Department of Justice guidelines, is unique to U.S. industry. The goal of the document is to help organizations understand whether they are in compliance with U.S. legal requirements. The second document, "Guide for the Study and Evaluation of an Organization's Environmental Management System," originally developed by the Institute for Environmental Auditing in the early 1980s, was revised by ASTM to be a national standard. The ASTM standard is supposed to be used to determine whether an entity has a quality EMS.

COALITION FOR ENVIRONMENTALLY RESPONSIBLE ECONOMIES' PRINCIPLES

Coalition for Environmentally Responsible Economies is a nonprofit membership organization composed of leading social investors, major environmental groups, public pensions, labor organizations, and public-interest groups. Their principles (formerly the Valdez Principles) are a model corporate code of environmental conduct. A company that endorses the CERES principles pledges to monitor and improve its efforts on:

- Protection of the biosphere
- Sustainable use of natural resources
- Reduction and disposal of wastes
- Energy conservation
- Risk reduction
- Safe products and services
- Environmental restoration
- Communication with the public
- Management commitment
- Audits
- Reports

Companies endorsing CERES principles must back up their pledge with specific information that is published in an annual report.

BRITISH STANDARDS INSTITUTE, "SPECIFICATIONS FOR ENVIRONMENTAL MANAGEMENT SYSTEMS"

British Standard 7750 (BS7750) was published in March 1992 and was the first major EMS standard model developed in the United Kingdom. The model served as the basic blueprint for the ISO 14001 (EMS) and, like the ISO 14001 model, offers a process for certification and registration. A number of companies have become BS 7750 certified, including a couple in North America.

Canadian Standards Association's CSA-2750, "Draft Guidelines for a Voluntary Environmental Management System"

Based on BS 7750, the CA-2750 is intended to provide general guidance to business, industry, and other organizations regarding environmental management systems. The CSA-2750 includes environmental management systems definitions, and principles, as well as fundamental procedures for implementing the model. It is expected that this standard is to be superseded by the ISO 14000 standard.

European Union's Eco-Management and Audit Scheme

Approved by the European Union in 1993, the Eco-Management and Auditing scheme is the European plan to help industry by prompting positive environmental management. The model provides for informing the public about the environmental performance of participating companies. It also focuses on performance instead of compliance with regulations. However, the model does provide for voluntary certification under a recognized EMS standard such as ISO 14000 or BS 7750. E.U. countries see the standards as a means of facilitating EMS certification and registration and as a way of demonstrating performance.

EPA's Code of Environmental Management Principles (CEMP)

The EPA has met the requirements of Executive Order 12856 by developing a draft Code of Environmental Management Principles for federal agencies. The code mirrors many of the principles that underlie EMS standards. Currently the draft CEMP is a benchmark containing organizational principles, infrastructures, and practices for a state-of-the-art EMS. As defined in the draft CEMP, a state-of-the-art EMS is one that assures that environmental performance will be considered as world-class or best-in-class by peers and stakeholders and will comply with the principle of the National Performance Review. The CEMP calls for the following:[3]

Management Commitment: The agency makes a written statement of management commitment to improve environmental performance by establishing policies that emphasize pollution prevention and the need to assure compliance with environmental requirements. Performance objectives include:

1. Obtain Management Support
   - Policy Development
   - Systems Integration
2. Environmental Stewardship and Sustainable Development

Compliance Assurance and Pollution Prevention: The agency implements proactive programs that aggressively identify and address potential compliance problem areas and utilize pollution prevention approaches to correct deficiencies and improve environmental performance. Performance objectives include:

1. Compliance Assurance
2. Emergency Preparedness
3. Pollution Prevention and Resources Conservation

Enabling Systems: The agency develops and implements the necessary measures to enable personnel to perform their functions consistent with regulatory requirements, agency environmental policies, and its overall mission. Performance objectives include:

1. Training
2. Structural Supports
3. Information Management, Communication, Documentation

Performance Accountability: The agency develops measures to address employees environmental performance and assure full accountability of environmental functions. Performance objectives include:

1. Responsibility, Authority, and Accountability
2. Performance Standards

Measurement and Improvement: The agency develops and implements a program to assess progress toward meeting its environmental goals and uses the results to improve environmental performance. Performance objectives include:

1. Evaluate Performance
   - Gather and Analyze Data
   - Institute Benchmarking
2. Continuous Improvement

ISO 14000

If you recall from your readings in Chapter 1, I provided an overview of ISO 14000 and very briefly explained the six standards. I'd like to go into more detail on these standards, now that you have a better understanding of the basic principles and other standards.

As you can see by reviewing Figure 2.2, ISO 14000 is actually six standards. These standards fall into two broad categories: organizational evaluation and product development evaluation. The organizational evaluation standards primarily deal with environmental and business management systems, including the Environmental Management System, Environmental Performance Evaluation, and the Environmental Auditing Standards. The second set of standards deal with the product development process and include Life Cycle Assessment, Environmental Labeling, and Product Standards. Let's now discuss each of these standards in a little more detail.

FIGURE 2.2   An overview of the ISO 14000 standards.

ENVIRONMENTAL MANAGEMENT SYSTEMS

The Environmental Management Systems (EMS) standard enables an organization to establish an effective management system as a foundation for environmental performance. The EMS provides the foundation for the entire environmental program and is essentially the cornerstone of the registration plan. The EMS contains three basic parts:

A written program. The EMS must be documented in either a separate manual or as parts of other documents such as quality plans or operations manuals. This document must portray senior management's absolute commitment to the highest-quality product with the lowest environmental impact and set forth written procedures to achieve this goal.

Education and training. If employees do not understand the goals of the senior management or have the training to implement these goals, they'll never be achievable.

An understanding of applicable laws and regulations. Essentially, this is an understanding of how to maintain compliance.

ENVIRONMENTAL PERFORMANCE EVALUATION

The Environmental Performance Evaluation (EPE) standards supplement the EMS standard. Initially, the EPE standard is intended to define the impact your organization has on the environment. This could be accomplished by conducting an inventory of those impacts such as solid waste generation, air emissions or hazardous waste disposal. Once a baseline of these generation rates is established, you can establish the reduction targets. Thus, the EPE is an ongoing evaluation by line employees responsible for the organization's environmental performance, as opposed to an independent audit by personnel apart from the organization.

ENVIRONMENTAL AUDITING

Like the ISO 9000 standards, ISO 14000 standards rely on auditing to assure the standards are being met. These audits may be conducted by people internal or external to your organization. Essentially, the audit evaluates the company environmental processes ranging from the inputs, including the raw materials, through the process, and to the outputs that are the products and the waste. The auditing standards are intended to:

- Guide organizations and auditors on the principles of environmental audits,
- Provide procedures for planning and performing an audit, and
- Address qualifications for auditors.

LIFE CYCLE ASSESSMENT

The standard in this category deals with a systematic way of examining environmental impacts of the inputs and outputs of a process that produces a product or service. In short, this standard forces us to look beyond the manufacturing processes to the environmental impacts of our products from operation and into disposal, reuse or recyclability. Although the concept is easy to understand, the application appears to be difficult. I see these standards being integrated into operations of an organization, addressing life cycles on a systems level rather than on a single-product level as the most effective method to apply this standard.

ENVIRONMENTAL LABELING

This standard provides principles for developing specific environmental claims on labels and provides purchasers of goods and services with a tool for choosing products based on environmental considerations. Essentially, the concept here is "consistency and accuracy in advertising". This could provide an advantage to "environmentally safe" products over "non-environmentally safe" products. The European Community (EC) has already adopted an Eco-label regulation that encourages manufacturers to reduce the environmental impacts of their products, while also informing the consumers about those impacts. I understand that these labeling standards currently are applied to detergents, paints, and paper.

ENVIRONMENTAL ASPECTS IN PRODUCT STANDARDS

This standard targets those organizations that create and set standards such as ANSI, ASTM, etc. It deals with considerations that should be taken when developing product standards to reduce environmental effects.

REGISTRATION AND CERTIFICATION

Once an EMS is fully functioning and an organization has passed an external audit with a registrar, then the organization's EMS can be ISO 14001 certified. Certification to ISO 14001 enhances an organization's image with the public and can help reduce the likelihood of fines and penalties. To customers and stakeholders, certification ensures that you're committed to being not just environmentally friendly, but environmentally proactive. Some of the other reasons to consider certification include:

- Fully satisfying the legal requirements
- Satisfying the contractual requirements
- Improving internal processes
- Reducing environmentally related operating costs
- Eliminating or reducing multiple assessments
- Meeting requirements of the marketplace

Certification can also play a role in some markets, especially the European markets where certification to ISO 14001 will be a legal requirement to sell products and services within that market. It can also become a contractual requirement of a customer or an important part of the supplier–customer relationship. Another advantage is that it can reduce the need for multiple on-site audits through increased trust with the regulatory agencies. This is because the discipline and process understanding required to achieve certification will be very challenging, therefore increasing the confidence of the regulatory agencies.

THE BASIC COMPONENTS OF THE ISO 14001 EMS

Now that you understand the overall ISO 14000 standards, let's further study the ISO 14001 EMS Standard. The 14001 model shown in Figure 2.3[4] provides a framework to develop and implement the EMS. The model identifies the basic steps that must take place to implement, operate, and maintain an effectively managed environmental program. Notice that this model is a continuous improvement type model in that it is an iterative process. It shows that all of your processes that affect the environment should be continually monitored and reviewed to identify opportunities to reduce their impact. There are five steps in this model.

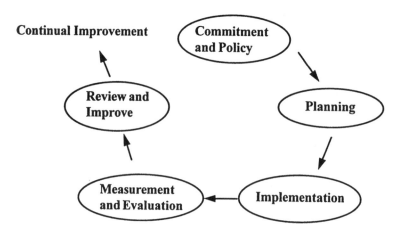

FIGURE 2.3   The EMS continuous improvement model.

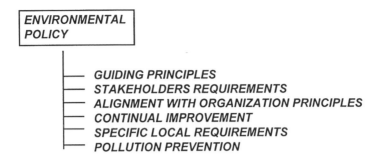

FIGURE 2.4   Environmental policy considerations.

ENVIRONMENTAL POLICY

The senior management of your organization is responsible for defining an environmental policy that is appropriate for your organization's activities, services, and products. Your environmental policy should be developed by senior management and must provide the vision or direction on where they want the EMS to focus. The direction is typically articulated through basic performance goals (Figure 2.4). As you develop your policy or guiding principles, I recommend you review other organizational documents and policies to assure your linkage and congruency with other organizational goals.

Your policy should indicate that your organization is committed to continual improvement, pollution prevention, and compliance with regulations. It should also provide a framework for setting and reviewing goals, targets and objectives. Additionally, management must assure that the policy is maintained, documented, and communicated to all employees and that it is made available to the public.

In summary, your policy should consider the following:

- Your organization's mission, vision, core values and beliefs
- Requirements of and communication with all stakeholders, including workers, customers, suppliers, local communities, and environmental regulators, as shown in Figure 2.5.
- Continual improvement, pollution prevention, and compliance
- Alignment with other organizational principles, such as quality and safety
- Any specific state or local conditions unique to that particular geographic location, product or process

Before developing the policy, an organizational environmental "checkup" should be conducted to evaluate the overall health of your program. The result of this checkup will determine where to place emphasis within your policy statement. Some things to consider when conducting this "checkup" are listed below. I also recommend you review the chapter on vulnerability assessments and strategic planning before developing your policy.

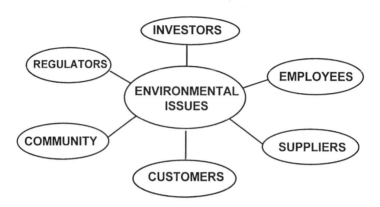

FIGURE 2.5   Environmental stakeholders.

- Reviewing your latest internal compliance audits. Are there any noncompliance areas?
- Reviewing your current environmental policies, procedures and checklists for congruency with laws as well as other organizational policies such as safety and quality improvement.
- Reviewing your procurement procedures to determine how they may impact purchasing environmentally friendly products.
- Checking your Emergency Planning and Community Right to Know program. Is it effective in collecting data? Can you make decisions based on that information? Is the public involved?
- Do you currently have an environmental policy? Was it ever officially approved by the organizational leadership. Is it current or does it reflect vision of your management to include the strategic planning factors like guiding principles? Who, by name and function is responsible for implementing the policy?
- Does your policy have you set goals and targets? Is there emphasis on continuous improvement?

PLANNING

The basic concept here is to develop a plan to implement the actions necessary to meet the policy you've developed in the step above. One of the key elements of planning is setting the targets and objectives in order to achieve these general goals set out in the policy statement. In setting objectives and targets, you should consider legal, technological, operational, and business requirements that impact all functions of your organization. The targets and objectives you set must be based on what is practicable and must also be measurable.

As you develop your plan, I recommend you include the rationale for the objectives and targets in the EMS manual. This is because once your EMS is in place, auditors will want to understand how the goals were established and be able to determine whether the organization achieved its environmental goals. The planning

FIGURE 2.6   Planning considerations.

section is broken down into four subcategories: environmental aspects, legal and other considerations, objectives and targets, and environmental management programs.

Environmental Aspects: The standards essentially ask you to identify environmental aspects of those activities, services, and products that can be controlled and that significantly impact the environment. Then you use this information when you set your goals and objectives, considering these impacts.

Legal and other Requirements: Your organization must be responsible for developing a procedure to identify, have access to, and understand the environmental requirements that you are required to follow.

Internal Performance Criteria: If external standards are not sufficient to define the performance of your organization as it relates to your objectives and targets, then you will be required to develop your own performance criteria.

Objectives and Targets: You must develop and document your objectives and targets at all functional levels within your organization to meet your environmental policies. These should follow and be consistent with your organization's TQM program and the commitments made to P2. These objectives and targets should be very specific and measurable over time. For example, "Achieve a 50% reduction in solid waste by the year 2000" is a specific objective and it's measurable. An ineffective objective is "Reduce solid waste."

Environmental Management Plans and Programs: Once you have established the targets and objectives, you develop a program to help attain those environmental goals and objectives. The plan you develop must clearly designate the responsible parties at all levels or functions within the organization. It must also address the resources required and time frame in which you expect to achieve your objectives. This plan may be updated as needed.

In Chapter 10, I have provided a detailed example of a management plan as it pertains to P2 efforts. This is a very detailed example for you to follow as you develop your own plan.

IMPLEMENTATION AND OPERATION

Senior management is responsible for developing the capabilities and support system required to achieve the policy discussed earlier. This includes such tasks as appointing personnel with defined roles, responsibilities, and authority for establishing the

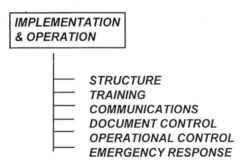

FIGURE 2.7 Implementation considerations.

EMS requirements and ensuring they are implemented and maintained. Implementing the EMS includes providing the required financial resources.

Implementation also includes making all employees aware of the policy and providing specialized training to those employees whose work has an actual or potential significant effect on the environment. The training should cover the potential consequences of departing from specified operating procedures and include emergency procedures for dealing with environmental incidents. The basic sections include:

Structure and Responsibility: Senior management must assign the roles and responsibilities and delegate the necessary authority to make sure there is sound management. This section also includes the necessary resources be provided — both human and fiscal resources. This section also requires that an individual is appointed to define the roles, responsibilities, and authority to make sure that EMS requirements are met and reported to senior management.

Training, Awareness and Competence: This section requires training be identified and provided to personnel within your organization whose work activities have the potential of harming the environment. The training procedures should cover:

- An overview and conformance with an organization's policy, procedures, and EMS
- Those work practices, procedures or activities that have the potential for impacting the environment; include discussions on how improved operational performance should lessen the impact to the environment.
- Clearly defined roles and responsibilities related to the requirements of the EMS, organizational policy, procedures, and emergency response.
- The outcome or result that could occur from not following the operational procedures outlined by the organization.

Communication and Reporting: You are required to establish, implement, and maintain procedures to enhance the flow of information to intra- and interdepartmental groups or functions, and to external agencies such as regulatory agencies or third-party registrar. These communication procedures are for such things as:

- Communicating management commitment and policies
- Raising stakeholder awareness
- Relationship building with customers, suppliers and regulatory agencies and external organizations

Environmental Management Systems Documentation: You must maintain information in any media that describe the core elements of your EMS, how they interact, and provide guidance to related documentation.

Document Control: Document control is an integral part of the EMS implementation. Management shall ensure that current versions of EMS and related procedures are available at all locations, any obsolete documents retained for legal and/or knowledge preservation purposes are suitably identified, and procedures and responsibilities are established for creating and modifying these documents. Also consider that documents can be located, reviewed periodically, updated as necessary, and approved by the authorized individual or authority. Any documents that are retained for legal or knowledge purposes must be properly identified as such. This section also discusses the need to have all documents legible, dated, and easily identifiable. Documents should be retained for a specific period. Procedures and responsibilities must be established for the creation and modification of various documents.

Operational Control: You must identify those organizational activities, processes, procedures, and operations that significantly impact the environment, as well as the objectives, goals, and policies related to these. These activities must be planned to assure they are implemented under specific conditions by developing and implementing documented procedures, specifying operating and performance criteria for all procedures, and establishing and maintaining procedures that are related to environmental aspects of the goods and services used by the organization. These procedures must be communicated to the particular suppliers and vendors as well.

Emergency Response: You are required to establish and maintain procedures to identify potential emergency situations. You are also required to develop and implement procedures for response to emergency situations and, where practicable, periodically test these procedures.

MEASURING AND CORRECTIVE ACTION

Your organization needs to develop, establish, and maintain procedures to measure, monitor, and evaluate those processes and practices that have an impact on the environment. This includes maintaining records on such things as calibration of measuring devices and records of the process, including all nonconformance and corrective actions. The records should be legible and traceable to each activity, stored in readily retrievable locations, and protected against damage, with established retention times for all records. Also, EMS internal audits should be carried out at defined intervals to evaluate conformance to policy and to check policy implementation.

The basic areas include:

Monitoring and Measurement. You are required to develop procedures that monitor and measure actual performance of those operations and activities that pose

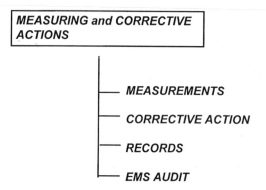

FIGURE 2.8   Measurement.

a potential threat to the environment. Information obtained from tracking the performance of your processes as well as outcome related to your organizational goals and objectives should be included. Any equipment you use to conduct the monitoring must be calibrated and maintained along with records that indicate this has been accomplished. Finally, you must also establish a procedure for periodically evaluating compliance to relevant environmental regulations. The information you gather from measurements should be analyzed and used to determine the success of the program and where to focus future activity or corrective actions.

Nonconformance and Corrective and Preventive Action. You are required to develop and maintain procedures that define the responsibility and authority for handling and investigating a nonconformance, taking action to lessen the impact of a nonconformance, and initiating correct action and preventive action. The corrective and preventive action should be appropriate to the magnitude of the problem and its associated impact on the environment. Any changes that are implemented as a result of corrective or preventive actions must be recorded in documented procedures.

Records. You are required to develop and maintain procedures for the identification, maintenance, and disposition of environmental records to include training records, audit results, and reviews. The records should be legible, identifiable, and traceable to the activity, product, or service it references. Records must be stored and maintained in such a way that they are readily retrievable and protected against damage, deterioration, or loss. Retention times must be established and recorded. Records must be maintained as appropriate, to demonstrate conformance to the requirements of the standard.

EMS Audits. You must develop a program and procedure for periodic EMS audits as outlined by ISO 14011.1 and 14012. In many ways this is a key section to the EMS standard. This is because auditing provides the organization with the ability to regulate itself and correct problems before they can be harmful to the environment. These audits are carried out to determine if your EMS conforms to the requirements of the standard and any requirements of the organization. They also determine if the EMS has been properly implemented and maintained and provides information on the results of the audit. This information must be communicated to management for periodic reviews.

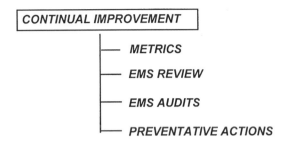

FIGURE 2.9   Continual improvement.

MANAGEMENT REVIEW

Senior management shall review the EMS to assure its continuing suitability, adequacy, and effectiveness. The review should happen continuously in an effort to improve the performance of the EMS. The EMS should address possible needs for changes in policy in light of audit findings, changing circumstances, and a commitment to continuous improvement. The review and changes shall be documented.

Review of the EMS. Senior management should conduct a review of the EMS at regular intervals to determine if it is successful. The review could include such things as:

- Review of objectives, targets, and environmental performance
- Review and root cause of findings from the EMS audit
- Evaluation of the EMS effectiveness
- Evaluation of the environmental policies

Preventive Actions and Continual Improvement. Essentially, this is the action the senior management takes to correct any discrepancies identified above. This includes checking that preventive actions are taken and that measurement systems are in place to assure problems will not recur. Continuous improvement is embodied within the entire EMS framework. It is the systematic approach to identify the problem, identify opportunities, implementing, measuring, comparing to targets.

RELATING EMS TO FUNDAMENTALS OF QUALITY SYSTEMS

The elements specified in EMS trace back to the fundamental three steps enumerated by Dr. Walter Shewart,[5] which constitute a dynamic scientific process of acquiring knowledge. The three steps in this process are *specification*, *production*, and *inspection*. Dr. Shewart explains these steps as:

- The specification of the quality of the thing wanted
- The production of things designed to meet the specification
- The inspection of the things produced to see whether they meet the specification

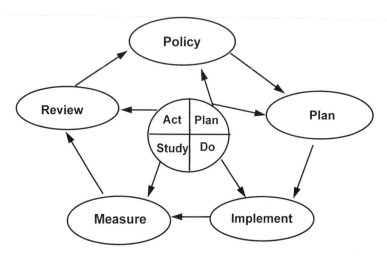

FIGURE 2.10   EMS cycle with the PDCA cycle.

These three steps were diagramatically represented as three arcs of a circle. In an idealized case where no changing of specifications is required, the production process will continue. In a changing environment, the specifications can be modified and the three-step process will emerge with new requirements.

Dr. W. Edward Deming called the above process the Shewart Cycle and later refined it into a four-step process[6] aimed at production. The four steps are:

- Design the product
- Make it; test it in the production line and in the laboratory
- Put it on the market
- Test it in service; find out what the user thinks of it and why the nonuser has not bought it

This four-step process is now popularly known as the "plan, do, check, act", or the PDCA Cycle.

The ISO 14001 EMS incorporates these four elements as follows:

- The "planning" stage is essentially the policy and planning stages of your EMS cycle.
- The "do" stage is the implementation and operation phase of your EMS.
- The "check or study stage" is checking and corrective action.
- The "act" stage is essentially the management review phase of your EMS.

As you can see, this cycle follows the continuous improvement mindset. It is a systematic or scientific approach to problem solving that includes gathering data and measuring trends in order to determine if progress is being made toward the goals, targets, and objectives you set in your policy and plan.

ISO 14000 AND THE BALDRIGE CRITERIA

In the United States, the Malcolm Baldrige National Quality Award was created to stimulate American companies to improve quality and productivity, while obtaining a competitive edge through increased profits. The seven criteria addressed in the criteria include:

- Leadership
- Information and analysis
- Strategic planning
- Human resource development and management
- Process management
- Business results
- Customer focus and satisfaction

Several of these criteria are addressed in various sections of the EMS as follows:

- ISO 14001 Section 4.1, Environmental Policy, requires commitment from management to define a framework for environmental issues, thus addressing Baldrige criterion 1, Leadership.
- ISO 14001 Section 4.2, Planning, deals with planning for environmental aspects, setting targets and objectives, and planning for legal and other requirements, thus addressing Baldrige criterion 3, Strategic Planning.
- Sections 4.3.1, 4.3.2, and 4.3.3 of the EMS address structure, responsibility, training, awareness, competence, and communication, which are in line with Baldrige criterion 4, Human Resource Development and Management.
- Section 4.4.2, Nonconformance and Corrective and Preventive Action, require procedures to identify nonconformance and corrective and preventive actions; and Section 4.4.4, Environmental Management System Audit, requires an audit. Both sections address Baldrige criterion 5, Process Management.
- Baldrige criterion 6, Business Results, can be obtained from Section 4.4.1, Monitoring and Measurement.

ISO 9000 AND ISO 14000

Subsequent to developing the 14000 standards, the ISO developed the 9000 series Quality Management Standards. The International Quality Standards (ISO 9000) were developed in 1987 to meet a need for common standards that would allow potential users to effectively evaluate products and services. It is used by 60 countries, including the United States, Canada, and Mexico. The ISO 9000 quality standards provide a framework for incorporating all the elements required to select, implement, manage, and monitor quality across a multinational environment. These

BALDRIGE CRITERIA	_EMS_
1. 0 LEADERSHIP	ENVIRONMENTAL POLICY
2.0 INFORMATION and ANALYSIS	IMPLEMENTATION & OPERATION
3.0 STRATEGIC PLANNING	PLANNING
4.0 HUMAN RESOURCES	IMPLEMENTATION
5.0 PROCESS MANAGEMENT OPERATION	IMPLEMENTATION & OPERATION
6.0 BUSINESS RESULTS	MONITORING and MEASUREMENT
7.0 CUSTOMER FOCUS & SATISFACTION	ALL

FIGURE 2.11   EMS analogous to the Baldrige criteria.

elements include traceability, documentation, audits, training, and management review. Some reasons organizations seek ISO 9000 registration include:

- Help compete in international markets
- Show commitment to quality
- Gain strategic advantage over noncertified competitors
- Meet customer requirements

The ISO 14000 standards are based on the ISO 9000 series and provide a framework by which corporate management can set and attain environmental goals that increase competitiveness and help ensure long-term productivity. Another attribute of ISO 14000 that can greatly enhance our environmental programs is the link to Total Quality Management (TQM). The TQM initiatives focus on the concepts of continuous improvement, strategic planning and setting goals and objectives. This environmental application of TQM is sometimes referred to as TQEM and can help environmental programs through:

- Systematically applying strategic plans that should link environmental concerns to organizational strategic and business plans.
- Creating focus by setting clearly defined goals and objectives. Part of this activity is identifying the primary OPR for each objective. This should expand the ownership of the environmental programs beyond that of just the environmental personnel.
- Creating accountability by measuring successes against those goals and objectives set in the strategic plan.

- Declining paperwork and duplication of tasks caused by redundant taskings, numerous reporting requirements, or various quality improvement teams.
- Reducing operating costs, thus creating a competitive advantage for having effective environmental programs.

The ISO 14000 series is analogous to the ISO 9000 series of standards for a quality assurance system. ISO 9000 has been widely adopted internationally. Even though the two quality standards are analogous, there are subtle differences. The ISO 9000 Quality Assurance Standards focus on the needs of the customer, while the ISO 14000 EMS standards focus on the needs of a broader range of interested parties and the societal need for environmental protection. In addition, the skills and disciplines needed for managing the environment and for conducting EMS assessments are distinct enough to warrant a separate set of quality standards.

CLOSING THOUGHTS

The environmental journey continues. We have moved from a period of regional and national compliance to a more global environmental view where environmental standards are becoming a necessity for international trade and commerce. Companies and organizations which impact the environment must begin to evaluate their existing management practices and current systems to determine their strengths and weaknesses. By reducing the weaknesses and inefficiencies, an organization can reduce unnecessary waste generation and potential liabilities. By following the guidelines of ISO 14000, as well as other EMS standards and codes, organizations can create more effective processes to reduce their impact on the environment. This step onto the EMS journey should not be taken lightly or without considerable thought and understanding. The resources for full implementation may be considerable and it will take time to fully implement an effective EMS. Although companies not paying attention to ISO implementation now will be paying catch-up later, this is a journey which should be fully planned and mapped before the first step is taken.

REFERENCES AND NOTES

1. A number of documents were reviewed to develop this section on Environmental Management Systems. In addition to those cited in Chapter 1, also see the following references for more information: a. ISO 14000 Offers Multiple Rewards, *Pollution Engineering,* June 1996; b. Block, M. R., ISO 14000, Universal Applications Level the Playing Field, *Waste Age,* 96–103, December, 1995; c. Rabac, G. and Stec, R., ISO 14000, The Groundwork for Environmental Management, p. 9, Perry Johnson Associates, Southfield, MI, 1995; d. Biggs, R.B. and Nestels, G. K., ISO 14000 — A Building Block for Redefining Environmental Protection and Moving Toward Sustainable Development, Weston Way newsletter, April/May 1995.
2. Nash, J. and Ehrenfeld, J., "Code Green", *Environment,* Jan./Feb. 1996.
3. Environmental Protection Agency document on the Code of Environmental Management Principles for Federal Agencies (undated).

4. A great many documents were reviewed to develop the section on ISO 14000. These include: a. ISO 14004 Draft International Standard, Environmental Management Systems — General Guidelines on Principles, Systems and Supporting Techniques, Aug. 1995. b. ISO 14001 Draft International Standard, Environmental Management Systems — Specifications with Guidance for Use, Aug. 1995, p. 12; c. ISO 14010, Draft International Standards, Guidelines for Environmental Auditing — General Principles, Aug. 1995; d. Kuhre, W. L., ISO 14001 Certification — *Environmental Management Systems,* 1995; e. Block, M. R., ISO 14000: Universal Applications Level the Playing Field, *Waste Age,* December 1995 p. 95-103; f. Scicchitano P., Managing the Environment with ISO 14000, *Quality Digest,* Nov. 1995, p. 43-46; g. Kinsella, J., ISO 14000 Standards for Environmental Management, Sept. 1994, See www.iso.14000.com; h. Crognale, G. G., Environmental Management: What ISO 14000 Brings to the Table, See www.iso.14000.com

5. Shewart, W., *Statistical Method from the Viewpoint of Quality Control,* Lancaster Press, Pittsburgh, PA, p. 45, 1939.

6. Deming, W. E., *Out of the Crisis,* MIT Center for Advanced Engineering Studies, Cambridge, Mass, p.180, 1986.

3 Leadership Implementation of the EMS

In Chapter 2, I introduced you to the EMS as outlined by ISO 14000. This, as with all environmental management systems, requires leadership commitment from all layers of the organization. To be truly effective, it must begin at the top, with senior management developing and then implementing an enviromental policy. Implementing your environmental policy, which is focused on preventing pollution and is bought-into by all employees, is vital to the success of reengineering your environmental systems.

An EMS requires you to develop systems which ensure all parts of your organization meet environmental goals without continuous oversight from an external environmental function. To achieve this you must have a formal environmental policy and standards that:

- Are communicated and understood by all employees;
- Ensure that environmental measures, objectives, and goals are developed and tracked;
- Monitor legal requirements and environmental compliance;
- Ensure that appropriate training is developed and implemented;
- Direct that adequate documentation exists and is controlled; and
- Develop and implement emergency preparedness and response.[1]

To effectively implement an EMS you'll begin changing the ethic of your organization though a coordinated and focused effort on the generation or handling of solid wastes, hazardous materials, and hazardous wastes. Assembling and organizing your team to attack these areas first is a critical aspect of building a successful EMS.

While environmental awareness is everyone's responsibility, some key players should be identified to plan, organize, initiate, and evaluate the EMS and P2 actions at each level within the organization. To maximize the effectiveness of an EMS and reduction efforts, a successful program will incorporate the elements shown in Figures 3.1 and 3.2 that include:

- An environmental steering group
- An ISO 14000/P2 working group

The Environmental Steering Group consists of the organizational senior managers who are responsible for developing and then directing the implementation of the environmental policy. The actual implementation of the policy occurs through a team or working group which I refer to as the ISO 14000/P2 Working Group. A coordinator may need to be appointed to help keep focus on the ISO 14000 and P2 activities.

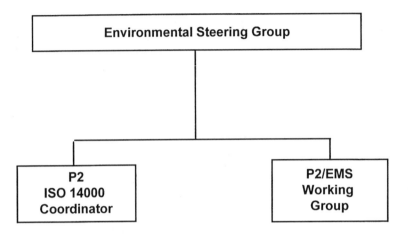

FIGURE 3.1   Structuring for success.

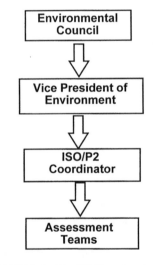

FIGURE 3.2   Establishing the structure.

THE ENVIRONMENTAL STEERING GROUP

This group of senior managers steers the environmental program by developing, implementing, and maintaining an EMS such as the one described in Chapter 2. As you recall, Figure 2.3 laid out the EMS continuous improvement model, which has five basic steps:

- Developing the organizational commitment and policy
- Planning to determine how you'll implement the policy and EMS

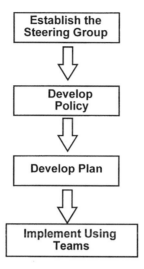

FIGURE 3.3   The sequential steps of the environmental steering group.

- Implementation of your plan, implementation the EMS
- Conducting measurements and other evaluations to see if you are meeting your policy
- Revision of your plan to continually improve

While this is a very simple model to understand conceptually, it is not detailed enough to use as a roadmap for implementation. I recommend the steps shown in Figure 3.3, which expands the first two stages of the EMS model; the "commitment and policy" stage, the "planning" stage to help you begin your implementation. These four steps will help you get your steering group operational and focused. The four steps are:

- First, you must establish the steering group;
- Then you develop your environmental policy;
- Next you formulate your plan by establishing goals and objectives;
- Then you implement the plan using teams.

By following these four steps to begin the EMS journey, you sequentially focus on specific tasks for your steering group so that it does not flounder in the implementation, bouncing from one topic to the next. Once these four steps are complete, move to the next stage of the EMS cycle, the "implementation" stage. I've combined diagrams 2.1 and 3.3 into one diagram (Figure 3.4), which I call the combined EMS model, to assist you to visualize where you are in the implementation process. Let's stay focused on the first stage, "commitment and policy," and conduct an in-depth review of the basic activities you will want to conduct for each of the four recommended steps.

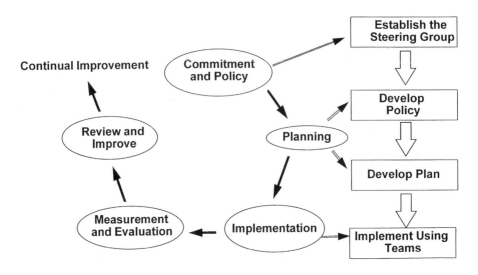

FIGURE 3.4   The combined EMS model.

ESTABLISHING THE STEERING GROUP

The first step in establishing the group is to identify the members. Membership should include senior managers from the major functional areas or companies within the organization. Special emphasis should be placed on operational units such as manufacturing and engineering, since these units typically produce the most waste and therefore play the major role in reducing the waste generated. Once you've established the membership, some general overview training on ISO 14000 and pollution prevention should be provided. As the steering group begins, monthly meetings are necessary while the policy is being developed and the goals are being identified. After that, I recommend that the steering group meet at least quarterly to review progress toward the goals and objectives which they set. They should also resolve any funding requests and follow up on any open items.

The products and services provided by the environmental steering group include:

- Developing explicit program scope and goals
- Developing the environmental policy that establishes the strategic direction
- Selecting and prioritizing P2 opportunities for implementation
- Matching prioritized opportunities and resources
- Chartering and monitoring the progress of P2 teams
- Tracking appropriate skills and training
- Authorizing resource expenditures
- Monitoring the progress of implemented opportunities

THE EMS AND P2 MANAGER

Depending on the size of your organization and complexity of your challenges, you may consider hiring one person specifically responsible to assist with the EMS development, act as the P2 project manager, and help establish and facilitate the environmental steering group. This EMS manager should work for the corporate director of environmental quality and has the task to openly communicate and coordinate the numerous actions taking place in the environmental arena. This person could also be in charge of benchmarking pollution prevention activities once the EMS is fully implemented, but one of the first activities this person could do is to assist with strategic environmental planning as discussed below.

DEVELOPING THE ENVIRONMENTAL POLICY

After the environmental steering group is established, and you've assessed your organizational health with respect to the environment, you'll then be ready to develop an environmental policy. This can be done through a detailed strategic planning exercise to help develop the policy, followed by conducting a gap analysis that will assist in developing the key focus areas, goals, and objectives which support and implement that policy. The policy is the vision of what senior management wants to accomplish and what will be different at the end of the improvements.

Recalling from Chapter 2, Figure 2.2, the environmental policy should include the organization's guiding principle; the stakeholder requirements; and indicate alignment with other organizational principles and values. The policy should include a commitment to continual improvement, pollution prevention, and maintaining environmental compliance.[2]

The organizational policy really provides the strategic directions or "vision" as to where senior leadership wants to guide the organization. It's critical that the environmental policy and goals align with other organizational principles, such as those laid out in quality, health, safety, or worker protection. Generally speaking, the policy statement should provide the vision for the environmental program. It should be the reason for existence of future efforts. The policy should be signed by the CEO or equivalent senior manager of your organization. This is critical to indicate that the policy is fully supported by the entire organization not just the environmental division. Some questions you can ask yourselves as you develop your policy include:

- What does the organization do?
- Who do you do it for? Who are the stakeholders?
- How do you do it?
- Why does the organization exist?
- Why do you want to prevent pollution or develop an EMS?
- How will we implement the policies we set forth?
- Who will implement the policies?
- How do you want the organization to be regarded?

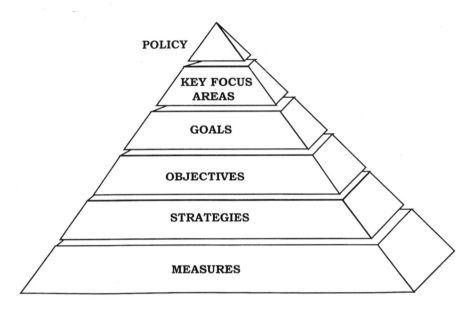

FIGURE 3.5   The EMS policy deployment pyramid.

Answering these questions may help you put your policy in perspective with other organizational policies. However, to really effectively develop and implement your policy you will probably need to use your strategic planning background.

STRATEGIC ENVIRONMENTAL PLANNING

Before we go further it's essential that you understand the basics of strategic planning. Understanding the concept and terms is essential for effective policy development, deployment, and implementation. The planning model I like to use is best described by a pyramid as shown in Figure 3.5, which I refer to as the "policy deployment pyramid." The various parts of the pyramid describe their logical relationships, beginning with a broad visionary policy statement and increasing in specificity through Key Focus Areas (KFAs), goals, objectives, strategies, and measures. Let's discuss each of these of these.

Policy Statement. As discussed in Chapter 2 and in detail in the section above, the policy statement is the heart of your EMS and P2 activities. It provides the vision as to where the organization is heading.

Key Focus Area (KFA). These essentially describe those areas upon which you may want to focus your energies to get the most results from limited resources. What are the major categories of customer requirements critical to the success of your organization? What are the major environmental requirements that are causing you the most problems or drain your resources the most? Chapter 5, the Vulnerability Analysis, provides screening tools to help you determine those key areas you may want to expend your energies on. Examples of key focus areas may include: cost reduction, personnel development, process improvement, or waste reduction by types of waste or media (hazardous waste, solid waste, air emissions), environmentally safe purchasing, public awareness, or maintaining compliance.

Goals. These are the specific areas where senior leadership wants you to place your effort subordinate to the KFAs. Some examples of goals would be to reduce solid waste or to reduce hazardous air emissions. Separate goals should be set for your reduction as well as your improvement efforts. ISO 14000 refers to this section as "objectives." Typically, goals are stated as phases such as:

- Promote sound waste management practices.
- Minimize the use of hazardous materials.
- Reduce air emissions to the lowest practical level.
- Improve operational performance through employee training.

Objectives. These are more specific than goals and typically force some type of action. Objectives must be measurable over time, as indicated by the examples below. Objectives should also have a "sun down" or completion date to them — this is the date when you expect to achieve your objective. ISO 14000 documents refer to these as "targets." Examples of objectives are:

- Reduce solid waste by 30% within 3 years
- All paper products purchased shall contain at least 20% post-consumer recycled content

Strategies. These are the actual steps you will go through to complete the objectives. In many cases, these include checklists, timelines, and other planning tools which describe how you'll achieve your objective. You need to develop the strategy that is most appropriate for your organization. The strategies become the plan or the beginning of the plan on how to achieve your objectives and goals.

Measuring Success. How do you determine if you are reaching your objectives? You need to measure. It may be necessary to first identify the amount of waste generated, then measure to ensure you are meeting your stated objectives.

Understanding this pyramid will help understand how the various portions of the EMS fit together prior to developing your policy and implementation plans. Then simply use this model to keep focused; working down the pyramid, first developing the policy, then the KFAs, goals, etc. Chapter 10 is a detailed example of part of an EMS manual and provides examples of the policy deployment pyramid for your review.

As you begin your EMS journey with policy development, I recommend that you involve any staff agencies that help conduct strategic planning for your organization and follow the model you are most comfortable with. If you do not have a model to follow, then follow the one I've provided below (Figure 3.6). This model is basically developed from traditional strategic planning for formulating policy, but it adds Hoshin planning elements used to align and review your strategy and plans.[3]

Assess your organizational values: Your organizational values define your organizational culture, including both personal and organizational beliefs. Most organizations

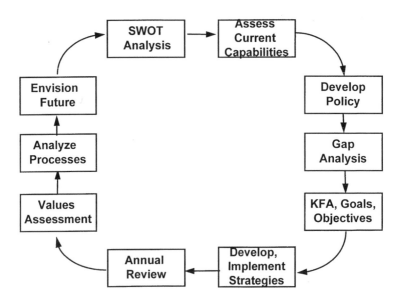

FIGURE 3.6 A strategic planning model.

have identified the values most important to them. These organizational values are integrated as the boundaries of any planning process and, of course, serve as a baseline for future decision-making. The reason you conduct a values assessment exercise is not to attempt to change an individual's values but to better understand the organization values and determine how they affect others. Once you have reviewed your values, then see if they align with your beliefs on how your organization should affect the environment.

Envision the future: Within this exercise you simply brainstorm as to where you want to be in the strategic nature. Once you have envisioned where you want to be, then, based on your current direction, you visualize where you will be in the future. Is your current direction leading toward your vision?

Conduct SWOT analysis: SWOT analysis is an exercise that helps you identify your "strengths, weaknesses, opportunity, and threats." This has also been called an "environmental scan." In this step you evaluate those factors, both external and internal, that affect your environmental management systems. When performing this analysis, it may be helpful to look at your environmental program as it is today, then look at it as you envision it in the future. Making a list of each of your strengths, weakness, opportunities, and threats will allow you to focus further discussion on each area as it affects your organizational vision — your policy.

Analyze your current capabilities: You'll need to begin an organizational vulnerability assessment concerning the environment. This is because you need to understand where your organization "stands" in terms of its environmental program before you develop your policy. Chapter 5 goes into greater detail on this subject of conducting a vulnerability assessment. Although a full assessment is not necessary before you develop your policy, a review of your organizational strengths and weaknesses is necessary. Essentially, you begin by analyzing your current processes and asking if they meet your expectations, and your stakeholders' and your customers' expectations. You may want to review your ability to stay within environmental compliance, or determine if you are expending too many resources on environmental issues. Perhaps the issue is public image. Each of these vulnerabilities affects your policy statement and further decision making points. Some common ways to conduct this review include: benchmarking, reviewing environmental audits and inspection reports, and conducting employee interviews. Then you envision how an environmentally responsible organization should function. Use this assessment as well as an "envisioning the future" type of exercise to help you create a baseline for where you are now.

These two strategic planning models are helpful in developing and then deploying your policy. But let's look at some example environmental policies that you could use as a template to develop your environmental policy.

- Our company is committed to continued excellence, leadership, and stewardship in protecting the environment. Environmental protection is a primary management responsibility as well as the responsibility of every employee and supplier of products to our organization. In keeping with this policy, our goal is to reduce waste and achieve minimal impact on the manmade and natural environment. Our basic guidelines include:
    - Environmental protection is a line responsibility and an important measure of employees performance. In addition, every employee is responsible for environmental protection in the same manner as personnel and worker safety.
    - Compliance with all federal, state, and local environmental regulations is paramount in all we do. Additionally, we are committed to reducing or eliminating the generation of waste. This will be a prime consideration in all future decision making.[4]
- At (your company), protecting the environment is one of our highest priorities and will be a major factor in future decision making. We are dedicated to continued improvement of our processes to prevent and reduce waste generation. At all times, we will maintain compliance with environmental laws, while we strive to reduce pollution at the source of generation.[5]
- The work of (your organization) has a major effect on the local environment. We are responsible for ensuring that we minimize the negative effects while we maximize the positive effects on the environment and work toward the achievement of zero emissions. In all of our activities, we will seek to:
    - Promote the conservation and sustainable use of natural and manmade materials
    - Minimize environmental pollution and waste
    - Build environmental concerns into all policies, programs, and services
    - Integrate environmental information into all levels of management
    - Achieve continuing improvements in environmental performance over and above regulatory and legislative requirements
    - Work in partnership with all stakeholders to promote waste reduction[6]

After you develop your policy, you'll need to present it to your employees, suppliers, customers, and other stakeholders. While senior managers and executives assign the priorities and set the stage for future efforts, the EMS must be institutionalized by the line employees. Your employees are the individuals who are responsible for the outcome of the program, so you must consider ways to motivate employees to get committed to the program.

Although some organizations use a lot of fanfare with their improvement programs, I don't see a need to do it with environmental and prevention based activities. Everybody I've ever worked with has been motivated to reduce unnecessary tasks and eliminate waste. My experience has shown that people do not need to be externally motivated with slogans, pitches, and other salesmanship. Just get them involved in the program. If they are involved with the development and agree with the basic concept, then they will have a significant impact on helping to reach your policy. Some suggestions include:

- Have the employees help define or develop the EMS goals and objectives.
- Have employees recommend waste reduction ideas and make certain that management responds to the ideas or provides feedback if the idea cannot be accomplished.
- Involve all levels of the organization on the assessment teams (which will be discussed in later chapters); also involve customers and suppliers.
- Let process owners have great input and final decision making authority on suggested improvements.
- Find ways to involve other stakeholders, such as suppliers and customers.

Once you've finalized the policy and employees are aware of it, the next steps are to put together a plan that will fulfill your environmental policy along with the goals that actually implement the policy.

DEVELOP YOUR PLANS; ESTABLISHING YOUR KFAs, GOALS, AND OBJECTIVES

Once you have completed your policy statement you will have a good idea of where the organization is heading in terms of environmental stewardship. Now it's time to develop your plan as to how you'll achieve your policy. This is stage 2 of the EMS model as well as the third step in the model I presented in Figure 3.4.

To begin this planning stage, take the policy and conduct a "gap analysis" against your existing organization processes. Essentially, you compare the current capabilities of those key processes to the expectations you set for yourselves in the policy statement development process. Are there gaps in your capabilities? These gaps become those critical issues for near and long-term goal setting. Study the gap. Is the gap between present and proposed capabilities reasonable? Would goals be achievable? If the gap is too large, perhaps the vision or policy is too aggressive. Perhaps it's not aggressive enough. Either way, the gap analysis is a good methodology for evaluating your policy statement and for determining your key focus areas and goals.

As you did with developing your policy, I suggest using the strategic planning process you use to develop your goals and objectives for other organizational initiatives to develop your EMS. In fact, the EMS should link to other strategic planning initiatives through the key focus areas and goals. As discussed earlier, P2 goals and objectives are integral to your EMS. The bottom line is that the P2 goals, objectives, and targets are the predominant part of your EMS in a proactive organization.

After you have developed your policy and goals, it's important to double check to ensure you have alignment with other organizational goals such as safety, quality, or personnel management to ensure you do not have conflicts within your organization. There are two basic types of alignment: vertical and horizontal. Vertical alignment I have been discussing. It is creating an understanding of your policy, goals, and objectives from the very top to the very bottom of the organization. Horizontal alignment is more difficult to achieve. It is ensuring the goals do not create conflict between functions such as engineering and manufacturing. The best way to check is to bounce it off employees who have various functions or to allow your P2/ISO 14000 team (discussed below) to review and make comments. After the key focus areas and goals are established by senior management, it's time to select the team to implement actions in an effort to reach those goals.

IMPLEMENTING YOUR EMS USING CROSS-FUNCTIONAL TEAMS

Now it's time to implement the plan you developed. There is no absolutely correct way to implement your EMS. It's simply a matter of following the plan you've laid out to achieve your policy and goals. Since the environmental business is very dynamic, with constantly changing requirements from regulators and other stakeholders, your implementation will need to be very dynamic. By focusing on allocating the appropriate resources and ensuring that your EMS goals are congruent with other organizational goals, you can increase the stability of your EMS. As I've discussed earlier, there is no need to try to implement all environmental goals at once. Pace yourself and focus on just one or two activities at a time. As you recall from our earlier discussions on ISO 14000, a major part of the implementation process includes such things as employee training, developing documentation records, and auditing your EMS. While these are extremely vital steps to fully managing your environmental programs, they are time consuming, expensive, and not necessarily focused on the end result of mitigating waste or your impact on the environment. That's why your P2 efforts are vital. Once you have your waste reduced to the most practical amount, then further develop your EMS. Do not develop the full EMS until you have reduced your waste streams to the most practical levels. Chapters 4 through 7 discuss P2 implementation, but now let's look at implementing the cross-functional teams.

The senior management team and the environmental staff are not the environmental team. The environmental team consists of every member of your corporation or organization. The employees must be part of the solution if you expect to have commitment and long-lasting results from your efforts. The environmental business belongs to all employees, just as just as worker safety issues belong to everybody in your organization. Effective P2 implementation requires a coordinated and focused effort on the part of all personnel involved in the generation or handling of solid wastes and hazardous materials and wastes. Senior management is responsible for providing the necessary manpower, money, materials, facilities, and information to ensure the EMS is properly implemented. The first step is applying the necessary manpower.

While pollution prevention is everyone's responsibility, key players and organizations must plan, organize, initiate, and evaluate pollution prevention actions. The development of Cross-Functional Teams (CFTs) or Integrated Product Teams (IPTs) is essential to the success of your initiatives because they involve all stakeholders with the environmental business. Typically, personnel are assigned to a team or project from the beginning to the project end, although many teams add people as technical expertise is needed and as the project progresses. It's also important to realize that CFTs are full-time membership, but they may be part-time work, meeting perhaps once a week, month, or on an as-needed basis. These teams should be formally chartered by the environmental steering board and should report to the board on their progress quarterly.

The P2 CFT typically proposes policy and recommends action to the environmental steering group. Depending on the size and complexity of the organization, the CFT may undertake the actions normally executed by the baseline and opportunity assessment teams, outlined in Chapters 5 and 6, but in many cases could also direct and review numerous assessment teams' activity. The CFT focuses on achieving specific goals, objectives, and targets identified in the organizations EMS such as the hazardous waste reduction or recycled material procurement goals. They report to the steering group on progress, initiatives, problems encountered, and other requirements. This step is crucial to the program's success as the overall working group cannot efficiently discuss, track, and implement initiatives to meet all of the organizations' pollution prevention goals. CFT personnel should stress the need for all personnel to be constantly looking for opportunities to reduce waste in their areas of expertise.

TEAM CHARTER

One problem that typically eludes this type of team is that of unclear objectives. Although the EMS policy may be clear in targeting the environmental goals, it may not be enough detail to keep this team on-track. To do this, a team charter may be required. The charter should be thought of as the contract between the environmental steering board and this team. It is a set of mutually agreed expectations of the team including the desired outcome. A team charter should contain some basic elements such as:

- A situation statement that essentially describes why the team was formed and the "as-is" state that the team is trying to improve.
- An objective statement that has a clear statement of the expected outcome, within a given time period.

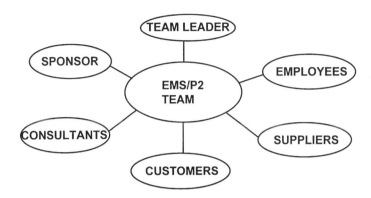

FIGURE 3.7   The EMS/P2 team stakeholders.

- A list of members and members' responsibilities.
- The criteria for determining the team success.
- A resource statement that defines funding, manpower, computers, or other resource issues.

TEAM MEMBERSHIP

The team should include a P2 Coordinator and personnel representing key functions that contribute to waste streams targeted for analysis. CFT membership should include those groups or departments that have significant operational or administrative interest in development of the program. Teams should be made up of five to seven shop chiefs and midlevel managers normally representing the facility maintenance, operations and procurement personnel. Other support elements necessary to implement change in operations to facilitate the reductions should also be represented. The team should include members who have direct knowledge of the processes that produce wastes or other harmful emissions, as well as technical advisors (Figure 3.7).

Typical functions represented on this team include facility maintenance, operations, procurement, safety and health professionals, manufacturing, public affairs, and perhaps an attorney. It may also be a good idea to have a union steward involved, as many pollution prevention opportunities involve personnel changes or operational changes that may effect job skills. Immediately after the team gets focused on a particular process or waste stream, ensure that the process experts are brought onto the team as technical consultants. Since nobody understands the process better, this person or these people are essential when gathering information on the "as-is" portion of the existing process. They can play a crucial role in implementing the new process through helping co-workers understand and accept the proposed changes.[7] Of course, some people are resistant to change or have the "not invented here syndrome," so by having the process owners become part of the solution you increase worker buy-in and thereby increase the chance of success.

While some organizations may use external consultants exclusively when dealing with data gathering and process analysis that yield improvement opportunities, I caution against this approach. I think that you get more help from consultants when they are integrated into a P2 team as technical advisors. The most successful programs I've observed are those where consultants are used to augment the staff or CFT. This reduces the likelihood of the "not-invented here" syndrome appearing and keeps the owners involved with the improvement ideas.

TEAM LEADER OR COORDINATOR

The team leader should be from the appropriate level for the company. As I recommended earlier, large organizations may want to consider hiring a single person to work both EMS and P2 issues; then this person could act as the team leader. The primary skills a leader needs are not technical skills, but rather organizational and communication skills. The coordinator also needs the ability to effectively listen, plus the ability to lead teams of 6 to 10 people, since the coordinator will be working at all levels of the organization. The coordinator needs to have a fairly broad understanding of the entire organization and to understand the disparate facility operations, and to have an ongoing interest in broadness and creative solutions to environmental problems. The corporate coordinator must be empowered with enough authority to keep the program on track as outlined in the EMS and organizational strategic plan. Depending on the amount and type of wastes this facility generates, P2 Champions should also be assigned for each plant or geographically separate area.

TEAM SPONSOR OR ADVISOR

The process owner should act as the team mentor. This person is the one individual who can change the process at the stroke of a pen. If there is not a single individual with that much control, then a senior manager could act as the team adviser. This person could assist the P2 CFT to pin down the exact process they are going to focus on first and could also act as a liaison between the environmental steering board and the cross-functional team while developing the charter and executing studies and recommendations. Dr. Lon Roberts[8] suggests having an Executive Sponsor when you plan on reengineering core business processes. I feel the same thought pertains to environmental processes as well. Some of the activities a sponsor or advisor could help with include:

- Assist in the development of the EMS and the P2 Strategic Plan, ensuring that there is direct linkage to other corporate goals and objectives
- Ensure that the P2 CFT stays focused on the EMS objectives
- Reduce parochialism and interdepartmental fighting
- Select the P2 team leader and ensure that the best members are put on the team
- Act as a sounding board and corporate mentor for the team. The sponsor should receive progress updates from the team to keep the team focused on chartered objectives and to keep the team informed of issues impacting their efforts

- Provide the necessary resources needed for the team to complete its tasks, which include:
  - A meeting place
  - Time
  - Supplies
  - Training
  - Relief from other duties or workload
- Track results of improvement efforts; follow-up on installed changes.

FACILITATOR

The facilitator is vital to the team's existence. Facilitators assist the team leader by focusing on the team dynamics; their main role is in how decisions are made vs. what decisions are made. The facilitator functions as an outside observer who works with the team leader to:

- Keep the team focused on objectives laid out in the charter
- Encourage constructive participation from all members
- Help use skills in group dynamics to improve the team's performance
- Properly structure the team meetings
- Teach team members to use various analytical tools

PRODUCTS AND SERVICES

The products and services provided by the EMS/P2 team include areas such as:

- Recommending overall program goals
- Monitoring overall progress toward achieving the goals and objectives
- Recommending implementation aspects of the policy
- Implementing a hazardous material identification and tracking system
- Implementing a waste-tracking system
- Prioritizing the waste streams, processes, or facility areas for assessment
- Selecting pollution prevention and opportunity assessment teams
- Providing training for assessment teams
- Establishing criteria for selecting options for implementation
- Conducting assessments
- Conducting technical and economic feasibility analyses of favorable options
- Selecting options for implementation
- Monitoring (and/or directing) implementation progress
- Monitoring performance of new option(s), once operational
- Ensuring accurate baseline calculations
- Classifying options generated by assessments
- Generating recommended solutions
- Selecting options for implementation
- Identifying funding mechanisms

- Documenting (and/or direct) implementation progress
- Documenting performance of new option(s), once operational
- Documenting overall progress toward achieving the goals and objectives

MEASURING AND EVALUATING PROGRESS

There is a management saying which essentially says: "What gets measured, gets improved." This is the final stage of the EMS implementation. Measurements taken over a period of time communicate the health of a process or activity.[9] Effective measurements present data that allow senior management to make meaningful "fact-based" decisions. These data-driven decisions that are based on quantitative measurements can help you break through to more productive ways to operate a process. You could measure many different parameters to determine the health of your processes, but because it is very time-consuming to collect and analyze the data, you'll want to make sure you measure the most important parameters. These are the measurements that are most closely linked to the objectives you have set. For example, let's suppose senior management set an objective to recycle 100% of all of your office paper. You could measure your total solid waste over time. This could be expensive, time-consuming, and it would not indicate the percentage of office paper being recycled. It would only show a reduction trend over time. A more appropriate measure may be a spot check or random sample of office waste on a regular basis. This measurement directly checks on the health of the activity. When you're developing measurement systems, you need to have open communications among all members of your team and with the members of the environmental steering group and process experts to ensure that you are measuring the right activities or outcomes. Some characteristics of a good measure include:

- It's meaningful to the user
- It's simple, understandable, logical, and repeatable
- It shows a trend
- It's clearly defined
- It's data that's economical to collect
- It's timely
- It drives appropriate action
- It tells how the organizational goals and objectives are being met through processes and activities

Some suggested indicators include things to measure, such as:

- Waste, rework, or other indicators of inefficiency
- Amount of waste disposed of over a period of time
- Amount of material purchased over time
- Amount of time (or people-hours) spent maintaining environmental compliance

Some of the tools you can use to help you display your data include trend charts, run charts, bar charts, and pareto charts, as discussed in Chapter 5.

CONTINUALLY IMPROVE

Once you begin measuring processes you will want to confirm that the actions have helped you achieve your objective rather than that some abnormality helped you achieve your objective. It's important to determine exactly why you met your objectives so that the solution is repeatable for similar processes and so that the original process stays under control over time. Some tools you can use to study the results include the pareto chart, control charts, and run charts. These help display and analyze data you gathered on checksheets. These tools are described in later chapters.

Maintaining the level of improvement can be achieved by integrating your efforts into your organization's daily operations and continually measuring, using the checksheets, run charts, and control charts. Some other steps you can take include publishing revised method and procedures, conducting training on the new process, creating periodic process reviews, and considering other areas that can be replicated.

After one target is met, it's time to determine what's to be done with the remaining problems or processes. The team should take this opportunity to evaluate the work accomplished, address remaining issues, and discuss with the steering group which other processes should be tackled next.

CLOSING THOUGHTS

Implementing your EMS includes establishing your steering group, developing your policy, and focusing on your goals. Next, it includes having a team in place readily trained and objectives and targets developed for the team to focus on. Once you have these in place it's time to begin your focus on P2 activities. In the next chapter, I develop the P2 cycle and show you how it integrates into the EMS cycle. Then, in the following three chapters, I go into great detail on how to implement these steps. The P2 implementation should be the first activities in your EMS journey. Focusing on reducing your waste before worrying about other things, such as registration or certification will pay dividends through reduced operating expenses and more stable systems.

REFERENCES AND NOTES

1. Ritter, C.A., ISO 14000 — A Corporate Perspective, Internet publication www.dep.state.pa.us/deputate/pollprev/iso14000/ritter.htm, p. 1.
2. Ritter, p. 3.
3. Holmes, S., The Quality Approach, 2nd edition, Sept. 1994, Air Force Quality Institute publication, p. 24–27.
4. Pingenot, J. and Voorhis, J., *The Pollution Prevention Assessment Manual for Texas Businesses,* prepared for the Texas Water Commission, 1995, p. 14–15.

5. Pollution Prevention for Federal Facilities, U.S. EPA Publication Number 300 B-94-007, 1994.
6. Adapted from EMAS Case Study, Hereford (England) City Council Pamphlet, undated.
7. Roberts, L., Process Reengineering, ASQC Quality Press, 1994, Milwaukee, p. 83.
8. Roberts, ibid., p. 78–79.
9. Holmes, ibid., p. 57–58.

4 An Overview of the Pollution Prevention Cycle

Our discussions so far have focused around your EMS. As I've stated many times, the most critical part of your environmental systems is your P2 program. Before you begin putting enormous effort into developing grandiose EMS documentation, you should focus your efforts on reducing the unnecessary waste being generated by less then optimally performing processes. To achieve this, I recommend you implement an effective P2 program. The incredible payback on many P2 initiatives will not only help generate support for your future efforts, it will also help reduce the amount of waste being generated and stabilize your environmental compliance activities.

Your organization needs to continually find ways to improve in order to remain competitive. By focusing on your operational inefficiencies, you may see that your processes may not be connected in the most logical manner, the wrong input materials are being used, or your internal processes are very unstable. The result is lower quality products, more scrap, more waste, and higher operating expenses.

Before I dive into detailed discussion of the various studies and steps of an effective P2 program, I would like to provide an overview of each of these steps and describe how they fit together into the EMS and the continuous improvement cycle we discussed earlier. The subsequent chapters go into further detail on each step of this cycle.

In general, the P2 cycle follows a "building block" approach for gathering and analyzing data. You use P2 studies to determine first how much waste, what type and form, and where the waste is generated. From that information you can conduct a detailed process analysis to develop a comprehensive set of prevention recommendations for implementation. By first identifying, in great detail, your waste generation prior to evaluating reduction opportunities, you have a better chance of attacking the "worst-first" waste streams. Prior to beginning detailed evaluations, you should first prioritize your facilities waste streams for evaluation and schedule the evaluations to be completed over a period of months. The evaluations are integral to the management strategy to be outlined by your EMS implementation plans.

THE P2 CONCEPTUAL MODEL

By reviewing the P2 conceptual model shown in Figure 4.1,[1] you see that the steps allow you to first seek to change processes to eliminate purchase and generation of undesirable "targeted" materials. Where this is not feasible or cannot eliminate waste, materials may be reused or recycled to reduce waste and purchase requirements. Goals to purchase environmentally preferable products through affirmative procurement programs help "close the recycling loop" by creating demand for recycled

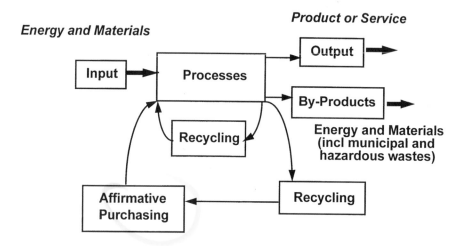

FIGURE 4.1   The P2 conceptual model.

products. Treatment of wastes and emissions control to reduce environmental impacts prior to disposal should be used only when other methods fail.

Typically, within the P2 hierarchy, the primary focus is on reduction or elimination at the source, followed by reuse, recycling, treatment, and finally disposal as a last resort. This is usually referred to as the "pillars of prevention", as shown in Figures 4.1 and 4.2. P2 efforts usually focus on reducing the use of hazardous materials, thus reducing the amount of hazardous waste and hazardous air pollutants generated by a facility. However, P2 efforts are not limited to just hazardous materials. There are great opportunities to reduce solid waste and air emissions and to find better business practices, as well as reducing energy and water consumption as well. These all lead to the goals of reducing operating expenses and minimizing environmental hassles.

Pollution prevention must be integrated into all activities across all functions to ensure that environmentally sound practices are in place. Although our discussion focuses on methods for reviewing existing operations to identify P2 opportunities, don't forget to pay equal attention to *new* operations and activities. It is during the planning and design phases that P2 activities will have the greatest impact on the total life cycle of a product, system, or facility. New facility construction, new industrial processes, and even projects to repair buildings and equipment, all offer opportunities to plan P2 into the project. Also, remember to consider P2 as a part of the environmental impact analysis process for all new projects and activities coming to your organization. By "designing-out" wastes and use of toxic materials from the start you will be fielding much simpler and "greener" systems.

THE P2 CYCLE

The P2 concept, as shown in Figures 4.1 and 4.2, is easily understood, but not always easily implemented. That's why I developed the P2 cycle that better describes the actual implementation of P2 efforts. The model is outlined in Figure 4.3. This model

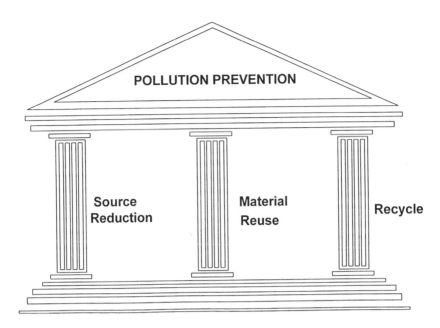

FIGURE 4.2   The pillars of prevention.

follows a systematic approach to plan, sequence, and implement improvement activities, using data you collected as the driver for decision making by senior management. Following this model allows you to collect and sort data and improvement opportunities into manageable elements. I have used or observed models with various steps ranging from 5 to 12 steps, all of which are powerful, but I developed this 8-step model because it breaks the steps into very understandable parts.

As you can see by careful observation, this process is continuous; by that I mean you must go back through each step, continually improving processes and reducing waste until the objectives set forth by senior leadership are achieved. Since this process is continuous, I feel it is better displayed in a circle rather than linear, as shown in Figure 4.4. I will refer to this as the P2 continuous improvement cycle.

This process fits nicely into the PDCA and the EMS models we've been discussing in our earlier chapters. As you can see in Figure 4.5, the P2 cycle mirrors the Plan-Do-Check-Act cycle.

Figure 4.6 shows how the P2 cycle connects into the ISO 14000 EMS process. Let's now discuss each of the steps in the P2 continuous improvement cycle.

STEP 1. IMPLEMENTING THE JOURNEY — POLICY AND OBJECTIVES

The policies, key result areas, goals, and objectives all come from your environmental strategic planning activities discussed in Chapter 3. These are the senior management strategies outlined in your EMS, plus any additional strategies and goals from other organizational strategic plans. This includes inputs from the corporate environmental steering group and may also include inputs from the state and federal

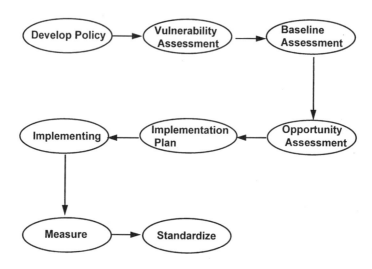

FIGURE 4.3   The P2 implementation steps.

FIGURE 4.4   The P2 continuous improvement cycle.

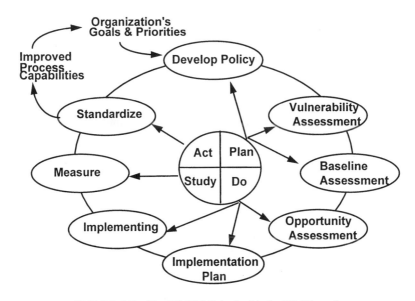

FIGURE 4.5 The P2 CIC linked with the PDCA cycle.

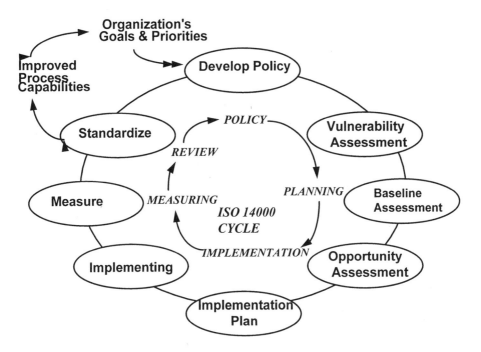

FIGURE 4.6 The P2 CIC linked with the ISO 14000 cycle.

regulatory communities and other stakeholders. In short, it is "Where the organizational leadership wants to go..." This area was discussed in great detail in Chapter 3 on leadership commitment.

STEP 2. VULNERABILITY ASSESSMENTS

Once you have received clear policy, and goals from your senior management it's time to begin the first of two steps in screening waste streams. Vulnerability assessments allow you to move from studying *all* waste streams to those *worst few* waste streams you'll want to concentrate on first. Ideally, all waste streams and operations would be evaluated quickly, but available time, people, and funds may require this to be done over a period of time. With this in mind, P2 evaluations should concentrate on the most important waste problems first, and then move on to the lower priority problems as time, people, and funding permit. Establishing the initial priority should be done with care. This is where the vulnerability analysis comes in.

Conducting the vulnerability assessment is usually a "paperwork" evaluation, although some interviews may be conducted. It is essentially a review of organizational policies, goals and objectives, plus waste disposal records, permits, and logs to determine which waste streams should be further evaluated. Based on these or other factors, the P2 manager and champions from the working group can prioritize the waste streams for further study or evaluation. Once scheduled, the process of evaluation will follow the same logical process for each waste stream, although some will be evaluated quickly and easily while others will require extensive study. For a detailed discussion of vulnerability assessments, see Chapter 5.

STEP 3. THE BASELINE ASSESSMENT

This is the second of two screening steps. In short, the baseline assessment determines "where you are now." When I use the term "baseline" with respect to P2, I'm simply referring to the amount of waste generated from a process, or material used in a process, at a given point in time. From these data you can judge your reduction progress and also decide exactly which waste streams are your worst waste producers. You need to determine how much waste you generate and from which waste stream, because:

- First, you'll want to focus your reduction efforts on the waste streams generating the "worst" wastes. In this context you define worst. It may be the most toxic waste, it may be volume, or weight, or the most expensive to dispose of. The leadership of your organization along with your help will define "worst" though the setting of the policy and goals.
- Second, you'll need to track reductions against a defined objective to determine if you are achieving your objective. For example, your objective may be a 50% reduction in solid waste over 3 years. A baseline survey is necessary to determine the amount of solid waste generated from all waste streams today so you have something to compare the reduction to and can judge if you're making progress toward your goals and objectives.

- The third reason the baseline survey is necessary is that it gets you and your P2 team to fully understand the who, what, when, where, and how's of every waste stream you assess. This detailed information is essential to making effective reduction decisions later on.

The result of the baseline assessment stage will be a database of waste generation. Various tools can be used to display the data, such as the pareto chart and other tools which will be discussed in Chapter 5. Detailed description of the baseline assessment is presented in Chapter 6.

STEP 4. OPPORTUNITY ASSESSMENTS (OAs)

The opportunity assessment (OA) is a detailed process analysis of those waste streams and processes you screened in steps 2 and 3. The purpose is to first fully understand the existing process, then identify the opportunities for waste reduction. The trick is to identify methods to reduce the amount of waste generated without negatively impacting the productivity of the process or quality of the product. Once you have finished the baseline assessment, you should have complete data on how much waste you generate and which processes generate the waste. The opportunity assessment focuses on waste reduction actions you must take to achieve your stated objectives. During this step you need to identify and verify the root cause of the waste generation through analyzing the process. By doing this analysis you'll avoid focusing on symptoms while identifying areas that need more information.

After the analysis is complete and you fully understand the existing waste generating process, then various techniques such as benchmarking or process reengineering can be conducted to identify possible process changes to reduce waste generation. I'll go into great detail on opportunity assessment in Chapter 7. Some of the typical steps include such things as:

- Understanding the existing process that generates the waste.
  - Process owners must flowchart the existing process.
  - Then the P2 working group can draw a simple process mass balance diagram to understand what happens to the materials used in the process.
  - Cause and effect analysis of the problem must be performed.
  - Potential causes must be analyzed to discover root causes.
- Asking the customer of the product you're producing what their actual requirements are. For example, If you're cleaning a part, how clean does it need to be?
- Brainstorming possible alternative methods: involve the shop personnel at this stage.
- If required, conducting a benchmarking study or literature search to determine if somebody else has found a more efficient process similar to yours.
- Developing a list of proposed options.
- Screening the options for technical, environmental, economic, and manpower constraints.

STEP 5. DEVELOPMENT OF THE IMPLEMENTATION STRATEGY

Once you've selected the proposed option, you need to develop a strategy for implementation. This step is essentially developing your action plan for implementation based on the information gathered during the previous steps. On the strategic planning pyramid I presented in Chapter 3, this was referred to as the "Implementation Strategy."

You'll want to plan actions that eliminate or reduce the root causes of waste based on the data you've gathered. Your action plan should include what, how, and when, and should identify the resources needed. Next, take a close look at your strategy and ask these questions:

- Are your methods feasible?
- Are they effective?
- What about cost benefits?

Answering these questions can help you develop a plan to implement those improvements. Other things you'll need are money and people; you'll need to get senior management commitment to provide the resources. To do this, you'll need to ensure your recommendation link to the EMS and the strategic plan for your organization. Chapter 8 goes into great detail on this.

STEP 6. IMPLEMENTING THE STRATEGY

Implementation is essentially the "do" portion of the "plan-do-check-act cycle." Once you have gained approval and resources from senior management, it's time to implement. If possible and practical, take action on a small scale at first. You may want to implement the improvements at one shop; then if it's effective, try other shops. Other things to consider in this step include:

- Scheduling the people to install or modify the equipment
- Installing the equipment or material or process change
- Training and involving the people involved with the change
- Establishing the measurement system

STEP 7. MEASURING THE RESULTS

Metrics and reporting are the key to continued success. Once you've implemented your improvements, the work isn't over. Now you must measure the amount of waste generated over time and track the trends to ensure that waste generation rates have gone down while other factors, such as product quality and defect rates, have remained constant or also improved. The trend should indicate that the action you took achieved the target. If it did, why? Are the results sustainable? Can you implement additional actions? At this point you may want to consider a control system such as Statistical Process Control (SPC) or other measurement system that will keep the waste-generating processes within the control limits you set.

Other ways to check on your progress include conducting environmental compliance audits and conducting an EMS audit. Audits are discussed in detail in Chapter 10.

STEP 8. TURNING THE WHEEL: STANDARDIZE THE SOLUTION

Now it's time to ensure that you maintain the improved level of performance by integrating your improvements into similar processes throughout your organization. If the improvement was only done on a small scale, then implement it on a larger scale. This step also allows you to plan what's to be done with the remaining waste streams and determine if additional teams or studies are required. This step allows you and your team the opportunity to review the work accomplished, address remaining issues, and evaluate effectiveness. Additionally, the team can review lessons learned in problem solving and interpersonal communications and group dynamics. Use this step to continue the process of implementation — go back to the P2 baseline assessment or vulnerability assessment and select another waste stream to reduce. Continue to do this until you have reached your reduction goals and objectives; then go through the cycle again, reevaluating your policy, goals, and objectives.

CLOSING THOUGHTS

Pollution prevention can greatly reduce your compliance-driven environmental workload by eliminating many hazardous materials and processes. It links directly into your EMS through your goals and objectives. Like the EMS process, P2 is a continuous improvement cycle with definitive steps. First your senior management develops the policy, or you use the policy from your EMS if it's developed. Then you conduct a vulnerability analysis to screen your waste streams down to those worst few. Third, you determine how much and what types of waste is generated by those worst few waste streams by conducting a baseline assessment. Then you analyze the waste generating activity in an effort to understand it and identify ways to reduce the waste without hurting the quality of the finished product. Implementation of your opportunity assessment strategy is the next step, followed by measuring your processes to determine if waste reduction is actually occurring. Once one waste stream is reduced, then you tackle another following the same process. The teamwork concepts discussed in Chapter 3 to implement your EMS efforts must also be applied with P2 implementation because you'll want to involve all process owners as a team focusing on alternative methods and processes to produce your products in an environmentally safer manner with fewer costs.

REFERENCES AND NOTES

1. Much of the discussion in this chapter was originally presented in two papers I coauthored: a. Dougherty, J. and Welch, T.E., Establishing a Pollution Prevention Program, in United States Air Force Environmental Quality Symposium Proceedings, Denver, CO, March 1993; b. Jacobs, J.M. and Welch, T.E., Implementing Pollution Prevention at Federal Facilities, in Air and Waste Management Association Conference Proceedings, Cincinnati, OH, June 1994.

5 The Vulnerability Analysis

How vulnerable is your organization to inefficient processes? to those practices which generate unnecessary waste through inefficiencies and improper input materials? How do you identify if they exist and determine which are the worst? That's where the vulnerability assessment come into play. The vulnerability assessment is simply a quick review of your processes and programs to determine into which areas you should invest additional study and analysis through baseline and opportunity assessment studies. If you have the resources to fully study all of your processes which generate waste then the vulnerability analysis phase becomes less important — but if you want to target your "worst-first," then this step helps identify those first few processes you should analyze. These vulnerable areas represent your "targets of opportunities" for improvements upon which to base those additional studies. This assessment can also be used with strategic planning to help develop your environmental policy, as we discussed in Chapter 3.

You may be asking the question "vulnerable to what?" Vulnerable to not meeting the policy, goals, and objectives you set forth in your EMS. One of your first steps is to determine what "vulnerable" means to your organization. If your EMS policy and goals are not clear on this you should consider some possible vulnerabilities such as:

- Waste and inefficiencies
- Noncompliance, and therefore vulnerability to fines and notice of violations
- Inefficient use of raw materials causing you to purchase more materials than necessary and, of course, generating more waste than necessary
- Having huge environmental compliance, health, and safety staffs because of the nature of the materials used in your processes
- Costs of implementation

In Chapter 3, I introduced you to the strategic planning process as shown in Figure 3.5. Let's take another look at it to refresh our memory (Figure 5.1).

This model describes the steps you can go through as you develop your policy, but it can also be followed as you evaluate your vulnerabilities. This occurs in two steps on the model: the gap analysis and the key focus areas, goals, and objectives step.

When you conduct the gap analysis you essentially compare where your environmental program is now to where you want to be to meet your environmental policy. If there are gaps in specific quantifiable areas, these could be the areas you focus your attention and improvement areas onto. For example, your policy may

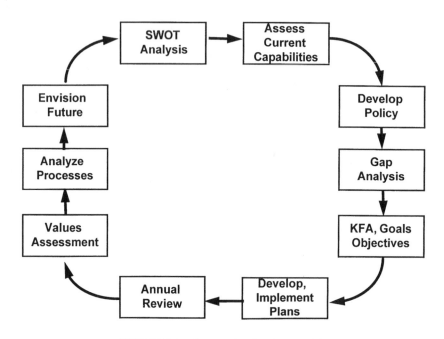

FIGURE 5.1   A strategic planning model.

have a statement about using environmentally safe products. You could conduct a quick analysis of your purchases to determine if you, in fact, are meeting your policy. The gap is those areas or purchases which need improvement to reach your stated policy. Perhaps you have a goal to purchase only paper products containing recycled content materials. Then you would check to determine if you are, in fact, meeting that goal.

Below I have developed some models to help you make this initial screen. I have provided a screen based on your organization goals, one based on risk, and finally a simple prioritization matrix. These models are very beneficial tools when working with a team, as they help you focus your attention on a few issues, thereby keeping your team on track. They are also helpful when you present your recommendation to senior management as you can show an analytical approach to the decision making and screening process. But before you begin great data collection exercises, let's firmly remember why you're doing the data collection — to have enough information to make good decisions on which waste streams to analyze first to reduce pollution and waste. Don't get "analysis paralysis"!

SCREENING WASTE STREAMS USING IMPACT TO HUMAN HEALTH AND THE ENVIRONMENT[1]

You can use this method to assess or consider the potential impacts a process or a waste stream has to human health and the environment under existing operational activities. The determination of risk at an organization is difficult because risk

WASTE STREAM NAME:

Questions	Level of Risk (High, Med, Low)
1. Compound (or Chemical) Characteristic a. Chemical Name: b. Exposure Limits: c. IDLH (Immediately Dangerous to Life of Health Level: d. Chemical and Physical Properties:	
2. Health Hazards a. Route of health hazard (inhalation, skin absorption, ingestion, and skin or eye contact): b. Symptoms: c. Target Organs: d. Level of Toxicity: e. Known Carcinogen?:	
3. Impact on achieving Environmental Policies and goals	
4. Environmental Hazardous: a. Type of exposure (air, land or water): b. Exposure routes (incinerator, surface container, underground tanks, injection wells, etc.) c. Observation of wildlife and plant health near the facility: d. Proximity to: Schools Churches Day Care Centers Residences Parks Restaurants Others e. Proximity to waterways	

FIGURE 5.2 Risk screening worksheet.

depends on constantly changing parameters defined by your operational procedures and precautions taken at your particular plant, line, or shop. Since determining in great detail the precise risk associated with each event is beyond the scope needed for this study, we will use a simplified method for determining potential risk for each waste stream.

Figure 5.2 contains questions to help you assess the potential risk. This worksheet will yield a high, medium, or low level of risk for each activity or process you consider. As you begin to analyze a waste stream, this method is somewhat rudimentary, but you will get more information on a particular process when you conduct your detailed opportunity assessment.

You may want to consult technical manuals such as the Material Safety Data Sheets (MSDS) or the *NIOSH Pocket Guide to Chemical Hazards*. These documents would help you complete information on the chemical compounds and health hazard. You also need to consider the environmental hazards to include both the ways in which the environment could be exposed to compounds in the waste stream and the potential risks associated with this exposure. As you consider environmental hazards, you must consider the known pathway of exposure. If there is no pathway, then the risk is quite low. Another important environmental hazard is to consider the impact the facility has on nearby plants and wildlife as well as impact on local gathering places. Finally, you should assess whether reducing this waste stream will have a great effect on achieving the policy and goals set forth by your senior management.

Answering the questions in Figure 5.2 may not give a direct indication of the level of risk, but it can provide meaningful data on the potential for risk. This should help you prioritize your waste streams to determine which one you want to focus your reduction efforts upon first.

SCREENING WASTE STREAMS USING WASTE STREAM PRIORITIZATION

If you have a general understanding of your major waste streams, then you can use this tool to prioritize the waste streams in an effort to determine which one should be reduced or eliminated first. The basic questions in Figure 5.3 should be assessed by relative weight and by relative rank. I recommend using the 1, 3, 5, 7, 9 scale to weigh your responses.

The number you assign to the relative weight column (W) will be dependent on your organizational goals and policy. Since the weight reflects the policy or goal, the relative weight of each criterion will remain constant for each type of waste stream. For example, the weight you place on each solid waste stream would be the same. The ranking (R) column will vary with each particular waste stream; it will reflect the need for waste reduction. The relative weight (W) should be multiplied by each rating (R) to fill in the column (R × W). Then simply sum the (R × W) column to calculate the sum of priority ratings. Comparing the waste streams sheets with one another will enable the team to give each waste stream a priority ranking. Some questions to ask while using this figure include:

- What are our policies, goals, and objectives?
- How high or low are your current disposal costs, and how high or low do you expect them to be in the future?
- Do current or proposed environmental regulations inhibit current processes?

WASTE STREAM NAME or ID:

Priority Rating Criteria	Relative Weight (W)	Rating (R)	R x W
1. Cost of Disposal			
2. Environmental Regulations			
3. Raw Material Costs			
4. Threat to workers, public, and the environment			
5. Processing Problems			
6. Amount of Waste Generated			
7. Toxicity of Waste Generated			
8. Residual Lifetimes			
9. Potential Liability			
10. Safety Hazard			
11. Impact on Policy			
Sum of Priority Ratings. SUM (R x W)			

FIGURE 5.3   Screening matrix using relative weights.

- Are raw material costs high now or projected to be in the future?
- Does the waste generated pose a direct, immediate, or perceived risk to workers human health or to the environment?
- Does the waste generated create complex problems for handling, shipping and disposing?
- How much volume is generated? Is it a problem now? In the future?
- Is the toxicity of the waste a problem?
- How long will the waste continue to display its hazardous characteristics? What is the residual life-time?
- Does disposal of this waste create the potential for long-term liability?
- Does using the raw material or handling the waste generated create a safety hazard for workers?

SCREENING USING ORGANIZATIONAL GOALS AND OBJECTIVES

It's very important in this early stage to determine what is important to your organization's leadership, because this will determine where to put further analysis and commitment on their part. Start by reviewing your ISO 14000 policy, goals, and objectives. These should help you determine where you should focus first. Is your focus on waste reduction, process improvement, saving manpower, saving money, or perhaps building the corporate image? The difference may appear subtle as you identify the possible targets of opportunities, but linkage to corporate strategies will become critical as you need resources to progress into further analysis and implementation. Before committing resources and effort its important to determine what

Key Process	In Compliance Y/N	Linkage to the EMS	Importance to the Organization	Resource Intensity	Needs Improvement
Compliance Costs					
Personnel Costs					
Worker Safety					
Expenses					
Occupational Health					

FIGURE 5.4 Screening using organizational goals and objectives.

is important to your bosses so that you can determine your data gathering plan. But before you begin great data collection exercises, let's firmly remember why you're doing the data collection — to have enough information to make good decisions on how to reduce pollution and waste. The concept of this simple matrix is to plug in a subjective value for each key business process as it relates to organizational goals. I recommend that you simply apply a scale of 1, 3, 5, 7, 9. The higher the value, the higher the likelihood of occurrence. The key process (row) with the highest numerical value when summed is the process you should focus on first (Figure 5.4).

After you have reviewed your organizational strategies and goals, objective, and targets, then you can begin the screening process. The questions below can be used to help you focus your attention on one of five areas: compliance, personnel, worker safety, operating costs, and corporate image. Although these areas are not mutually exclusive of one another and the end result of the effort may be the same — reducing waste, it's vital to focus your effort on an area that senior management has dictated as being important.

- Reduce the effort you put into complying with environmental laws. You may decide to focus your P2 efforts on those few waste streams that are taking your staff's time and energy and driving you out of compliance.
- Evaluate your compliance with existing laws. Are you currently or have you been out-of-compliance with any environmental laws and regulations? If so, do you have a full understanding of which laws are violated, and the root cause of the violation? Some places you may be able to find information and questions you can ask are listed below.
  - Have you done an environmental compliance audit? If so, review the findings from the audit to determine if there are opportunities for improvement.
  - Do you develop numerous management plans and waste permits? Include such plans as hazardous waste management plans, waste minimization plans, spill response plans, waste discharge permits, disposal permits. These plans are very difficult and labor intensive to develop and maintain.

WASTE STREAM NAME:

EVENT	PROBABILITY of OCCURRENCE	IMPACT	SUM
Non Compliance	5	7	12
Personnel Costs	5	9	14
Worker Safety	5	3	8
P2 Initiative	7	9	16
Public Image	3	7	10

FIGURE 5.5   Screening using probability of occurrence.

- Reduce the personnel costs to the organization. The focus here is to reduce waste from the processes which are most labor intensive. Some questions to ask yourself when considering how to rank these include:
  - Do you have a large staff of personnel working to maintain compliance? If so, which media? hazardous waste, air, water, solid waste, hazardous material?
  - Do you contract out to consulting firms for a great deal of your compliance work?
  - Do you see a tangible benefit to your company through environmental stewardship?
- Hazardous material and safety issues. The focus in the area would be on those hazardous materials which drive worker safety and personnel protection issues, and on applying P2 initiatives which reduce the impact from using those materials. Some questions to ask yourself include:
  - Do you have a large staff of personnel who help protect workers from hazardous materials and safety-related issues?
  - Do you spend significant resources on PPE, spill response equipment, and hazardous material storage equipment and facilities?
- Reduce your expenses. The focus here would be on reducing the disposal costs through application of P2 initiative to reduce waste generation.

SCREENING USING PROBABILITY OF OCCURRENCE[2]

As I discussed earlier, you probably can't evaluate all of your waste streams at one time. So assessing the risk of each waste stream is another method you can use to screen your processes. I have presented a simplified risk assessment model for you to use (Figure 5.5). Although this model is incomplete for extremely complex processes, it does serve as a framework on which to screen your processes. For detailed risk analysis you may want to consult another text.

Risk assessment typically involves three components: the event, the probability of that event occurring, and the impact should that event occur. I've used this context to develop this example.

- First you identify the event that could occur. This event could be a notice of noncompliance, unnecessary waste generation, excess expenditure of manpower or whatever.
- Second, you identify the probability of that event occurring. This will be a subjective value. I recommend you simply apply a weighted scale such as the 1, 3, 5, 7, 9 scale I suggested earlier. The higher the value, the higher the probability of occurrence. Don't put a lot of effort deciding on the probability of occurrence, just apply the weight.
- Third, you identify the impact to your organization should the event occur. Again, for simplicity, I recommend you use a weighted scale. The higher the value, the more damaging the impact is on the organization.
- Another factor to consider is the climate of your organization when it comes to risk. If your organization or perhaps your boss or process owner is risk adverse, then you'll also want to consider that among your weighing factors.
- Another potential consideration is to weigh potential impact to a organization or process if the P2 initiative fails after it is implemented. Is the organizational impact high?

SCREENING USING THE PRIORITIZATION MATRIX

Using the prioritization matrix is one tool to help you "rack and stack" your waste stream or processes. This tool is simply a way to multivote or "weigh" many different parameters for a single process as shown below. To multivote, you need many people, such as your ISO 14000/P2 team. The team silently votes using a weighted scale. Each member votes on the relative importance of each parameter such as those I've shown below. The parameters you chose may be different than the parameters or characteristics I have shown in the example below:

- Compliance with regulatory requirements
- Energy usage
- Overall cost of managing the waste
- Threat to human health
- Threat to the environment
- Overall quantity of the waste generated
- The P2 potential or ease of implementation
- Cost to make a change
- Value of any "recycled" material
- Improvements to the quality of the end product
- Reduction (or increases) in person-hours
- Reduction in the amount of waste generated

Each team member applies a simple numeric value to each parameter for each process. Using the higher value to indicate "bad" or "undesirable" events, then

WASTE STREAM NAME

Parameter	Team Member 1	Team Member 2	Team Member 3	Team Member 4	Sum
Compliance Costs					
Personnel Costs					
Worker Safety					
Expenses					
Raw Material Costs					

FIGURE 5.6   Team members prioritization matrix.

simply sum the numbers for each process. The process with the highest numeric value should be dealt with first (Figure 5.6).

TOOLS FOR SCREENING USING EXISTING DATA

Typically, many organizations have a lot of data which is required to be gathered to maintain compliance with various regulatory requirements. These data can be sorted using a variety of tools, including the fish-bone diagram, the pareto chart, the bar chart, and the pie chart. By analyzing available data it may become clear which area you should focus on. The figures and discussions below will provide you with an understanding of how to use some of these tools.

THE CAUSE–EFFECT DIAGRAM

This diagram is also known as the fish-bone diagram and is best applied when you want to graphically display possible causes of a specific problem for which you have very little data and need to identify a starting point to begin analysis. Essentially the diagram illustrates the relationship between a given outcome and all the factors that influence this outcome. Figure 5.7 shows the possible factors influencing the outcome of P2, while Figure 5.8 displays possible causes of source reduction. To use this tool you typically state the problem to be overcome or state the objective to be accomplished and place this in the box on the right side of the diagram — this is the effect. Then you list the major categories influencing the effect being studied. These are typically the 4 Ms: "manpower, method, machinery, and material" or the 4 Ps: "policies, people, procedures, and plant"[3] although any heading may be used. From this basic starting point you begin brainstorming, asking why, or using other idea-generating techniques to place possible causes on the diagram.

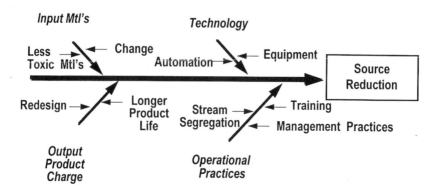

FIGURE 5.7   The cause-effect diagram.

THE PARETO CHART

A pareto chart is a type of bar chart which helps you visualize the relative "size" of one problem to another. The notion behind the pareto analysis is the 80, 20 distribution — 20% of the problems have 80% of the impact. The 20% are the vital few problems where you should focus your attention. Separating the data in this way helps you prioritize the relative importance of priority of one problem over another. To use this tool you need to have some data. You can either use existing data or gather the information you need. Then you group the data by consistent units of measures such as pounds, man hours, dollars, etc. Next, you simply create a bar chart with the frequency of occurrence on the left vertical axis and categories of problems on the X or horizontal axis. As you plot the data you order the categories according to their frequency of occurrence (how many), not their classification (what kind) in descending order from left to right. I also like to use a right vertical axis to measure the cumulative percentage of total occurrences summed over all the categories, as shown in Figure 5.8.

This figure displays the results of an internal organizational assessment of hazardous waste streams. This example shows that 80% of the hazardous waste generated at this particular facility was generated by three waste streams. It's obvious that most of the organization's problems are caused by unused or expired shelf-life hazardous material and paint related hazardous waste. This initial analysis indicates that additional data gathering should take place in this area. A second chart could be drawn to show the detailed analysis of any one of the three waste streams. This is sometimes called the "nested pareto" analysis.

PIE CHARTS

These are another great method for displaying the relative size of one problem to another. We're all familiar with pie charts, so I won't bore you with the details of constructing one. Figure 5.9 is an example of "nested" pie charts displaying data on types of noncompliance violations and the failure analysis data on one slice of the pie. I also plotted this information in Figures 5.10 and 5.11 to display it in pareto charts from the discussion above. As you can see by analyzing these diagrams, pie

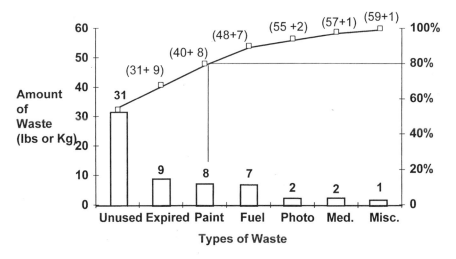

FIGURE 5.8   A pareto chart.

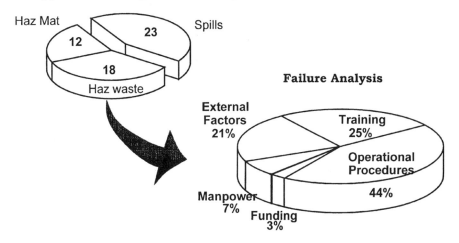

FIGURE 5.9   Nested pie chart examples — displays non-compliance violations and failure analysis.

charts are better used if you have only a few categories, while the pareto analysis is more appropriate with significant data and numerous categories.

Bar Charts

Bar charts continue to be one of the simplest and most effective tools for displaying data. Example 5.11 graphically displays non-compliance trends based on internal assessments. You can easily see that hazardous materials and hazardous waste are the leading problems in this example without the help of the pareto 80/20 analysis.

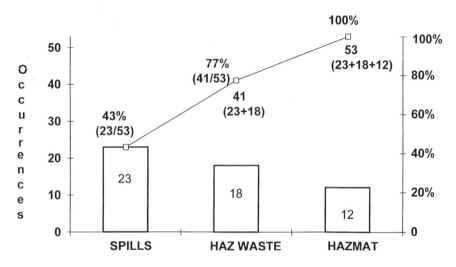

FIGURE 5.10   Pareto chart example — using a Pareto chart to display the results of the failure analysis from Figure 5.9.

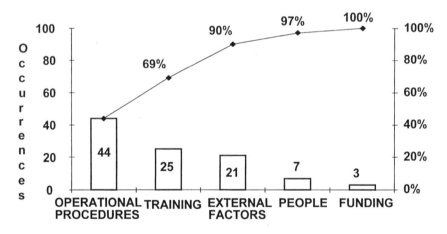

FIGURE 5.11   Pareto chart example (Cont.).

TOOLS FOR GENERATING IDEAS

Our vulnerability analysis so far has assumed you have some data to analyze. Let's now assume you have no data. What tools can you use to determine your vulnerabilities when you have little or no data? Essentially, your only option is idea-generating tools in which you solicit comments from all levels of your organization in a very structured method which focuses on the process of identifying problems instead of personalities. Let's discuss some of these methods.

BRAINSTORMING

Brainstorming is used to develop a list of potential ideas or vulnerabilities. We've all used brainstorming before, so I won't go into the details. Some variations include structured brainstorming, where you solicit ideas one at a time from each member of the group. There is also silent brainstorming, which works well when some members may feel intimidated by other members of the group. In this method everyone writes his ideas on sticky-back notes which are placed on the board for all members to see. Once you have a list of ideas, then you can narrow them down through discussion or through some of the other tools such as multivoting or the use of the affinity diagram, or nominal group technique.

THE "FIVE WHYS"

This method allows you to get to the root cause of a problem by asking why until the answer does not yield any more useful information. This usually takes asking why a few times — hence the name "five whys."

MENTAL IMAGING

In this method you first visualize the problem solved in the ideal state and then visualize the obstacles you must overcome to achieve this state or goal. This process is very much like the gap analysis; your objective is to identify the main obstacles to achieving your goal.

THEMATIC CONTENT ANALYSIS

This tool is an analytical technique used to help summarize and categorize data from surveys, questionnaires, and interviews. You simply list each question from the survey or questionnaire and attempt to summarize the recurring themes from the responses. This tool is very helpful when you are trying to obtain many options on where you should place your efforts.

There are many other tools, such as tree diagrams and pairwise ranking to help you use information more effectively. Using and understanding these tools will help your team focus on the most important issues affecting your environmental management systems.

CLOSING THOUGHTS

If you're convinced that you need to focus your efforts on the "worst" or most vulnerable processes, but if you're not sure what those are, then run through the tools above to help you sort information. Remember, if you choose to conduct a vulnerability analysis, the sole purpose is to focus your attention onto the most important waste streams for prevention opportunities. By doing this "worst-first" analysis you can decide with fact-based decision making which waste streams, processes, or programs to tackle/improve first. There are numerous analytical tools which will be helpful in determining where to focus your efforts. In the next phase, the baseline assessment, you begin analyzing those few processes which make you most vulnerable.

REFERENCES AND NOTES

1. Pingenot, J. and Voorhis, J., The Pollution Prevention Assement Manual for Texas Businesses, prepared for the Texas Water Commission, 1995, p. 34–35.
2. Roberts, L., *Process Reengineering,* ASQC Quality Press, Milwaukee, 1994, p. 81. This and other texts on business process reengineering provide insightful process analysis. This is very closely related to the pollution prevention opportunity assessment phase where you essentially reengineer a process to reduce your impact on the manmade or natural environment. You may consider reviewing some material on reengineering prior to conducting your vulnerability assessments.
3. *The Quality Approach,* U.S. Air Force Handbook 90-502, August 1, 1996, p. 60–80. This text is one of numerous handbooks, pamplets, and other publications which describe the use of tools and techniques to help make teams and decision-making more effective. I recommend you review these techiques as you begin your ISO 14000/P2 journey.

6 The Baseline Assessment

Once you have completed the vulnerability assessment to determine exactly which waste streams you are going to analyze first, then you can begin the baseline assessment. A baseline is simply a measure of the amount of material purchased, or waste generated, from a specific process, during a specific period of time. It's usually measured in weight or volume. Once you have determined the baseline amount generated from a waste stream, you use the data to help you further establish which waste streams you will reduce first and then track the status of your reduction initiatives in meeting your P2 goals or objectives. The baseline becomes the basis for calculating your progress toward the reduction goal or objective.

For example, suppose your EMS objective is to reduce hazardous waste by 10% annually. You'll need to know the amount of waste you currently generate in order to achieve your reduction objective. The baseline assessment may show that the amount of hazardous waste generated from a specific waste stream is 250 pounds in 1994. You can then track the success of your prevention efforts against the baseline on a trend chart such as the one shown in Figure 6.1. Additionally, by conducting baseline surveys on numerous waste streams you'll gather information on the waste streams and the processes that will help you determine where to focus your attention for those reduction efforts. You'll determine which processes generate the most waste, or the most toxic waste. With this information, you'll then be able to further determine which waste stream to reduce first.

Typically, baseline information includes the information below for each waste stream.

- The quantity of material purchased or generated during a specific time period
- The name of the substance
- The unit cost of purchase and/or disposal
- The using or generating process

We'll discuss this in more detail later in the chapter.

WHY BOTHER CONDUCTING A BASELINE ASSESSMENT?

Have you ever heard the saying: "If you don't know where you're going you'll probably end up someplace else and any road will take you there?" This saying is applicable to the baseline assessment because the baseline survey establishes "where you are" before you can plan your journey. The baseline, along with the vulnerability analysis discussed in Chapter 5, is the starting point for the road map that you'll

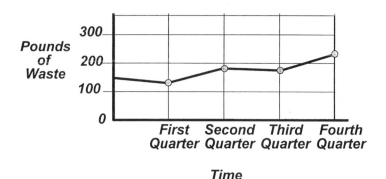

FIGURE 6.1   Trend chart showing waste generated over time.

follow to achieve your EMS objectives. They provide data for decision-making on where to focus your efforts. In this manner you achieve those objectives set forth in the EMS — those that are most important to your organization.

The baseline assessment helps you determine, in great detail, how much and the type of waste being generated by each process. Once you know the "worst" generators of waste you can then determine where to focus your efforts. After all, you have limited resources so why make reductions to small waste streams? Why not focus on the worst first?

First you established the baseline of waste generated from each process and by the waste stream as a whole, i.e., all hazardous waste or all solid waste. Then you can measure the progress of your P2 initiatives. You do this by comparing the quantities of toxic substances used or waste generated in the time period before and after implementation of the P2 initiative. Hopefully your trend chart will show a decrease in waste generated, as it does in Figure 6.1. I'll discuss this further in Chapter 7; let's now discuss the data-gathering methods.

TWO APPROACHES

There are at least two ways to gather data on your facility's operations: one approach focuses on the input end or toxic substance use, while the other approach focuses on the output end or waste streams. Both approaches have their strengths and weaknesses, as discussed below.

FOCUS ON THE INPUT — THE MATERIALS AND THE PROCESSES

The "input" approach starts with a review of available material purchasing logs and other hazardous material information. Once you have gathered these data you use them to develop usage or generation trends. The concept here is that by reducing hazardous material usage within a process you not only reduce the hazardous waste generated, but you also reduce the worker protection requirements. The basic methodology for an "input" approach involves:

- Searching logs, procurement records, MSDSs, and EPCRA records for chemical usage data and hazardous waste and solid waste generation data
- Totaling these data for each type of toxic material, or by process
- Identifying a limited number of processes that use the largest amounts of the targeted substances or generate the largest or most toxic waste streams, for further study and development of pollution prevention opportunities

The advantage of this approach is that it minimizes the labor required to generate a baseline of information. The disadvantage is that it does not provide a full picture of all processes that may yield worthwhile pollution prevention opportunities. It evaluates only those processes that use toxic or perhaps hazardous material, regardless of whether they actually generate waste or are preventing you from reaching your reduction goals.

THE WASTE GENERATION APPROACH

In this "output" approach you begin with identifying every activity that produces wastes, then perform detailed analysis to determine the toxicity or the amount of waste generated. The methodology for this study requires:

- Identifying all activities that use targeted substances or generate waste;
- Fully documenting each process, the material it uses, and wastes it generates; and
- Summing all activities' material usage and waste generation quantities to obtain the totals for each program component (air, water, solid waste).

The advantage of this approach is that it provides a comprehensive picture of all waste generating processes and provides a complete and detailed foundation for development of pollution prevention opportunities. The disadvantage is that it requires extensive effort and knowledge of all processes.

THE COMBINED APPROACH

My experience gained while conducting assessments has shown that data from the two approaches will seldom agree. Using the "input" approach, it is difficult to gather complete data using records that were never designed with environmental data collection needs in mind. For the "output" approach, detailed records of material use and waste generation at the process level are seldom available, and estimating will always introduce errors. So, in the end, a compromise between the two approaches usually provides the best solution. You may want to consider quantifying the total material used for each waste-generating process you're analyzing using the input approach. Then select the activities you identify to use the most targeted toxic chemicals or hazardous material. From there, use the output approach to identify which of these waste streams identified generate the most wastes, and perform a process analysis on the selected activities. This process analysis methodology is laid out in steps 4 through 10 of the pollution prevention cycle and will be discussed in

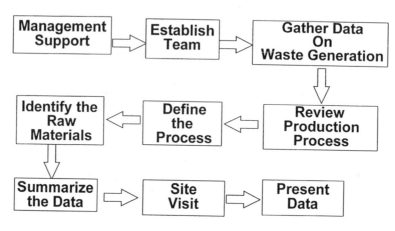

FIGURE 6.2   The baseline assessment sequential steps.

Chapters 7 and 8. If the selected activities don't yield enough reduction to meet the goals you set in your EMS, then do another iteration of activity selection and analysis.

Your approach should consider the nature of the activities or process you plan to evaluate. Very complex processes probably need further analysis than do simple processes. You also need to consider which provides the most useful information while making the best use of limited data and personnel resources. Since you have already screened your processes down to a handful, through the vulnerability analysis, then you may have the resources available to conduct detailed baseline assessment on that handful of processes. You will need to analyze a number of processes to identify enough material and waste reduction opportunities to meet your EMS goals and objectives. Also, the identification of these opportunities still requires detailed analysis of the inputs, outputs, and costs for each process before selection of alternatives and final implementation takes place.

THE STEPS TO CONDUCTING AN ASSESSMENT[1]

The baseline assessment stage is conducted after the vulnerability assessment that screens all processes down to those few which you want to focus on first. The baseline should consist of iterative steps of gathering data from records reviews, then analyzing and sorting the data, followed by conducting site visits to verify data, fill the data gaps, and interview process owners. These steps are shown graphically in Figure 6.2. After you have the data, you then summarize it as shown in Figures 6.4 through 6.8. Let's discuss each of the steps in detail.

STEP 1. MANAGEMENT SUPPORT

A key element in implementing the baseline is receiving management support through the commitment of the necessary resources to support the assessment activities.

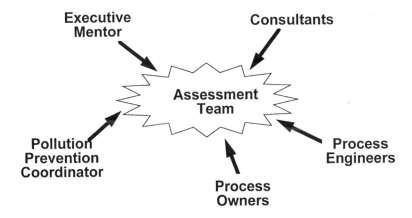

FIGURE 6.3 A typical baseline assessment team.

Management has communicated their support through development and publication of the organization policy, an EMS, and also by appointing a person to facilitate the P2 activities. Now they need to commit people to conduct the baseline assessment. You could get the commitment to your proposed plan through a detailed presentation to the environmental steering group. The presentation should cover the need for conducting the assessment along with a detailed outline of the steps your assessment team will follow. Also include the expected costs and benefits. I also recommend that you draft a team charter that should include information such as we discussed in Chapter 3 on setting up the P2/ISO 14000 cross-functional team.

STEP 2. THE BASELINE TEAM

The baseline assessment team (Figure 6.3) should be led by the EMS/P2 coordinator. The size and composition of the team will vary, depending on the size and complexity of your waste-generating processes. In general, the team should be multidisciplinary, composed of individuals with substantial technical, operational, and communication skills. The baseline team may be the same as the P2/ISO 14000 cross-functional team or it may be separate. Either way, membership should include personnel from production, maintenance, and process engineering and could also include budgeting, purchasing, research and development, and environmental management. One person from senior management could act as team mentor.

One of the first tasks the team should do is to review the team charter, the organization's EMS, and any data gathered and analyzed during the vulnerability assessment phase. The review should allow all team members to gain an equal understanding of their roles and responsibilities; to understand the organizational priorities; and learn the worst waste-generating activities. Figure 6.4 is provided to assist you as you develop your team. It is a checklist you can use to assess your team members' general knowledge of the baseline assessment phase and determine if some general or introductory training is required.

How well do your team members understand each of these areas?			
Members	Role in Understanding Your Facility	Role in Changing Your Facility	Role in Maintaining Your P2/ISO1400 Program
Process Owner			
Management			
P2 Coordinator			
Finance/Accounting /Purchasing			
Research and Development			
Process Engineering			
Quality Control			
Environmental			
Sales & Marketing			
Safety & Health			
Production			
Facilities Maintenance			
Material/Inventory Control			

FIGURE 6.4   The baseline team self-assessment checklist.

The second task the team should complete is to develop shop questionnaires based on those provided in Chapter 11. These questionnaires should be developed for those shops you currently plan to visit and may be expanded as the project continues. Once these questionnaires are written, they should be forwarded to the appropriate shop to complete.

STEP 3. GATHERING WASTE-GENERATION DATA

The data-gathering step begins with document research and is followed by an analysis of these documents. If you have not done the steps recommended in the vulnerability analysis stage, it's time to do these steps. Before you run out of your office and begin conducting site visits, I recommend that you review your existing documentation. For most waste streams, you probably already have a great amount of information. Then, you can use the site visit to verify the information and to fill any gaps. Figure 6.5 is provided to help you summarize information you will need from each process you analyze. Figure 6.6 may also be helpful in describing the product or service output from each process you analyze.

I like to use a checksheet such as the one shown in Figure 6.7 to help keep organized. This diagram is helpful to keep track of documents and other information you gather for each waste stream or process.

Operation Type: ___ Continuous ___ Discrete ___ Batch or Semi-Batch ___ Other						
Document	Status					
	Complete? Y/N	Current? Y/N	Last Revision	Used in this Analysis?	Document Number	Location
Process Flow Diagram						
Material and Energy Balance						
Design						
Operating						
Flow Measurements						
Stream						
Analyses/Assays						
Process Description						
Operating Manuals						
Equipment List						
Equipment Specifications						
Piping and Instrument Drawings						
Plot and Elevation Plan (s)						
Work Flow Diagrams						
Permit/Permit Applications						
Batch Sheet (s)						
Material Application Sheets						
Inventory Records						
Product Composition Sheets						
Operator Logs						
Production Schedules						

FIGURE 6.5   Information needed from each process.

The next two diagrams I use while conducting baseline assessments are shown in Figures 6.8 and 6.9. These diagrams allow me to summarize information on the input material for each process I'm studying as well as to summarize the output and waste material for that same process.

Each of these forms should be modified to meet your needs. If you have not fully decided where to begin your baseline assessment, I recommend that your data collection should begin with hazardous waste information followed by other media as listed.

Attribute	Description		
	Product or Service	Product or Service	Product or Service
Product or Service Name/ID			
Major Components of Product or Service			
Total Annual Production			
Number of Units Sold			
Number of Units Reworked			
Number of Units Scrapped			
Average Unit Production Cost			
Average Sale Price			
Annual Revenues			
Shipping Mode			
Shipping Container Size and Type			
On-site Storage Mode			
Containers Returnable (Y/N)			
Product Shelf Life			
Scrap Rate			

FIGURE 6.6   Product or service summary.

Hazardous Waste. Let's start our data gathering by looking at hazardous waste generation. This is a good place to begin because many organizations track these data quite well and this is where most opportunities for improvement tend to be. Also, most of the information has been already collected under various regulatory compliance programs. Information on hazardous waste manifested off-site, permits, waste effluent characteristics, and information on the chemicals used in your facility are a great starting point. Identify the waste streams (e.g., spent cleaning solvent, sludge, chemical baths) generated from each process, operation, or practice (including maintenance). For each waste stream, describe the quantities generated, the physical or chemical composition, and the associated handling/treatment/disposal activities.

- Gather the following documents:
  - Hazardous waste manifests
  - Annual hazardous waste generator reports
  - Lists of types and amounts of hazardous material used by production crews
  - Permits for hazmat, hazwaste, and air pollution
  - Industrial wastewater flow records and reports
  - Industrial wastewater and storm water permits
  - Lab reports characterization data

Shop Name:
Point of Contact
Street Address /Phone Number
Products Produced:
Process Description:
List the operations at this shop/building:
Provide a rough estimate of production volume per month (e.g., the number of circuit boards produced.
How many people work in this shop/building? How many hours is it operational?

What are the major waste streams?
What media do they affect?
How much waste is generated in each?
Has the shop/building already implemented any waste reduction or recycling techniques?
What are the major waste streams?
What media do they effect?
How much waste is generated to each media?
What is the waste toxicity or concentration?

Process ID:
Process Title:
Components of Concern:
Consumption Rate:
Purchase price of the material:
Delivery Mode:
Storage Method:
House Keeping Practices:
Shelf Life Issues:
Container Issues:
Known substitutes:
Alternate Suppliers:

FIGURE 6.7 Process information.

- NDPES monitoring reports
- Emission inventories
- Biennial hazardous waste reports
- Environmental audit reports
- Permits and permit applications
- Review the above listed documents in an attempt to accurately describe the amount and types of waste generated from various processes within your facility. Give each process that generates hazardous waste a unique process identifier or name, and place the information onto a spreadsheet or diagram as discussed above.
- Plot the waste generated by the various types, such as paints, solvents, and cleaners, on a pareto chart.
- Plot the amount of hazardous material generated over time in a "run chart" to track the trend of waste generation. During the baseline, you will need to do this for every waste stream you analyze. See Figure 6.1 for an example of a run chart.
- Do you have, or have you had any "Notices of Violations" or other environmental speeding tickets? If so, review these to determine the root cause of the problem.

Attribute	Description		
	Toxic Substance No.	Toxic Substance No.	Toxic Substance No..
Input/Substance Name/ID			
Source/Supplier			
Important Physical Characteristics (solid, liquid, etc.)			
Important Chemical Characteristics (caustic, corrosive, etc.)			
Function in Process (feedstock, reagent, intermediary, etc.)			
Final Destination (consumed, process waste, reused, etc.)			
Total Annual Consumption Rate			
Total Annual Cost			
Delivery Mode (pipeline, tank car, truck, etc.)			
Shipping Container Size and Type (55 gal drum, 100 lb. paper bag, tank, etc.)			
Storage Mode (outdoor, warehouse, underground tank, etc.)			
Transfer Mode (pump, forklift, conveyor, pneumatic transport, etc.)			
Empty Container Disposal Management (clean & recycle, landfill, return to supplier, etc.)			
Supplier would:			
- accept expired material? (Y/N)			
- accept shipping containers? (Y/N)			
- revise expiration dates? (Y/N)			

FIGURE 6.8   Input material summary for toxic substance use.

- Do you track the amount of hours it takes to manage hazardous waste and hazardous material compliance? If so, review these to determine the root cause.
- Review any engineering data such as design reports to determine specific waste and material stream characteristics.
  - Process flow diagrams
  - Material energy balance (by design and by operations)
  - Flow or amount measurements
  - Process descriptions
  - Operating manuals

Attribute	Description		
	Wastestream No.	Wastestream No.	Wastestream No.
Wastestream Name/ID			
Source/Origin			
Major Wastestream Components			
Wastestream Physical Properties			
Wastestream Chemical Properties			
Total Annual Generation Rate			
Method of Collection			
Method of Management or Disposal (sanitary landfill, hazardous waste landfill. POTW discharge. on-site recycling. etc.)			
Unit Disposal Container (55 gal drum. tank, lab pack, etc.)			
Disposal Costs			
Unit Cost			
Total Annual Cost			
Regulatory Compliance			
Potential Liability			
Safety Hazard			
Source Reduction Potential			
Minimization Potential			
Potential By-Product Recovery			
Potential On-Site Recycling			

FIGURE 6.9   Waste stream summary.

- Equipment lists and specifications
- Piping and instrument diagrams
- Work flow diagrams
- Material application diagrams
- Production schedules

Air Quality. Air quality assessments can be very detailed and, in my opinion, "painful" because of the amount of iterative analysis you must do. Air quality goes hand-in-hand with hazardous material and waste assessments because most air emissions are from hazardous materials used in processes. Some considerations for data gathering include those listed below, as well as those discussed for hazardous waste.

- Does your facility have air permits? If so, gather the permits and review the hazardous emissions listed within the permits.
- List the parameters that state, local, and federal regulators require you to track.
- Develop a matrix of the amount of waste generated.
- Draw a "trend chart" that describes the amount of waste generated over time.
- List the equipment and processes covered under these permits.
- Emissions monitoring is very difficult; list the requirements for your facility
  - List or estimate the number of man-hours you put into emission monitoring.
  - Estimate costs of outside consultants.
  - List the cost of maintaining compliance.

Wastewater. Typically, wastewater discharges are clearly dictated under your discharge permits. Areas to consider include the following:

- What type of wastewater does your facility have? (List whether it's sanitary, storm, industrial, or other type of wastewater).
- List the wastewater discharge permits you may have. What parameters, in quantity, type, etc., do you discharge?
- Do you have a storm water permit or any type of storm water characterization program? Do you know how much waste and what type of waste is in your runoff?
- Does your facility use or discharge any of the priority pollutants?
- If so, have you analyzed your effluent or runoff for the priority pollutants?
- Do you have regulatory reports, such as National Pollution Discharge Elimination (NPDES) permits or SARA Title III reports. These provide information on the volume, composition, and toxicity of wastewater discharged.

Financial and Other Related Information. This information may not be needed for the baseline survey, but it may be needed to convince your senior management to approve a particular project. If it's available, gather it.

- Raw material costs
- Water, wastewater, storm water costs
- HW disposal costs
- Solid waste disposal costs
- O&M costs for those operations that could possibly be altered

STEP 4: PRODUCTION QUANTITIES AND GENERATION INDEX

The amount of waste generated and hazardous material used is directly linked to production quantities. You may need to develop a generation index to determine actual quantity of waste reduced after taking into account changes in production levels. To do this, identify the major products produced (e.g., printed circuit board refurbishing, plating of components) or services provided (e.g., jet/automobile/equipment engine repair, dry-cleaning, aircraft painting) by your organization. Identify the amount/numbers of product produced (e.g., square feet of printed circuit board, gallons of plating solution) or service provided (e.g., number of various engines repaired, number of articles of clothing dry-cleaned) for each process or operation. This information can usually be obtained from:

- Material application diagrams
- Material safety data sheets
- Product and raw material inventory records
- Operator data logs
- Operating procedures
- Production schedules
- Production maintenance records
- Job control records
- Cost accounting reports

The result of this step should include an understanding of the type, amount, and toxicity of waste generated so an index can be developed to normalize the baseline of waste generated against the production activity. It is easy to see a reduction trend in waste generated if the amount produced is also reduced. Having an index normalizes the waste generation with production. If you have analyzed many waste streams, and you need to compare the quantify generated to other waste streams, then you'll need to develop a waste generation index. This is necessary because, in general, waste generation or toxic substance usage is directly dependent on a facility's production rate or level of activity.

The actual index you use will depend on the size and complexity of your operations, and the number of variables that must be considered. You may use a single index for all operations or perhaps use a different index for each separate waste stream, process, operation, or toxic usage. The unit, process, or production index is typically a ratio of production levels or level of activity for a facility from the two time periods of interest. Let me provide some examples on how to calculate this type of index.

- The paint shop used 5000 gallons of paint this year and 6250 gallons in the previous year. The majority of the waste generated and toxic substances used is associated with the painting processes. The "production" index for this facility and its processes is 0.8 (5000/6250).
- A dry-cleaner cleaned 3300 garments in this year, and 2700 garments in the previous year. The majority of the waste generated and toxic substances used is related to the dry-cleaning process. The "process" index for this dry-cleaner is 1.22 (3300/2700).

The activity index can be a ratio of another activity conducted within the facility that is not necessarily associated with production or business activity. As an example:

- A shop refurbishing composite parts for aircraft uses acetone to clean the repair molds. The molds are cleaned on an as-needed basis that is not necessarily a function of the production rate. Operators cleaned 520 molds this year, but only cleaned 200 molds in the previous year. The "activity" index for this acetone usage is 2.6 (520/200).

Obviously, each of the examples provided above includes only 2 years. You can also calculate trends or develop other indexes such as waste toxicity, disposal costs, etc. I caution that you can easily get into "analysis paralysis" if you are not careful. All you want to achieve with the indexing analysis is to rationally decide which of the waste streams you analyzed are the best ones to eliminate first. Once you have calculated the appropriate index for your waste streams, you can use it to both measure the progress for P2 efforts that have already been conducted, or for any efforts that you will be undertaking in the future. These indexes should be reported in any discussion or analysis of past or completed P2 activities.

STEP 5. UNDERSTANDING THE PROCESS

This is an optional step in the baseline survey phase as it will be predominantly used in the next phase, the opportunity assessment. If you have the information available, and you're pretty sure you will be pursuing reduction efforts on this waste stream, then it's a good thing to do now. For each product or service, delineate the current operations processes and practices used and, if possible, break these operations, processes, and practices down into individual steps. For each discrete step, compile this information on a worksheet. This information can usually be obtained from:

- Process flow diagrams
- Operating manuals and process descriptions
- Facility or equipment layout drawings
- Typical process flow diagrams, input-output models, piping diagrams, etc.
- Material balance sheets for production and pollution control processes
- Equipment lists for those processes that generate wastes
- Equipment layout
- Material logistics flow maps
- Product composition and batch sheets
- Operating procedures

STEP 6. IDENTIFYING THE TOXIC SUBSTANCES

This is a step that could also be conducted during the next phase, but if time permits I suggest that you attempt to identify the hazardous and toxic materials usage for the process you are studying. Later on, during the opportunity assessment phase, you'll be asking why the particular material is being used, so it is a good idea to identify the locations and type used now. Having hazardous material makes you vulnerable to a host of environmental and health-related laws and regulations, ranging from OSHA hazardous communication to spill response and planning exercised under EPCRA. Reducing even one hazardous material reduces your vulnerability to increased operational costs.

You want to be able to describe the materials used as inputs for each process, operation, or practice. For each substance, describe the quantities used, the primary function of this material (feed stock, catalyst, intermediary) and final destination (consumed in product, process waste, reused). This information can be summarized on Figure 6.8 and can usually be obtained from:

- Form R (EPCRA Section 313)
- Inventory and purchasing records
- Operating manuals and process descriptions
- Production or job control records
- Operator data logs
- Material Safety Data Sheets

STEP 7. SUMMARIZING

As I've discussed earlier, the P2 surveys require collecting and analyzing much data. These data are usually summarized on worksheets such as those shown in Figures 6.3 through 6.9. For each process or waste stream you have reviewed, I recommend that you document it on the information worksheet, such as the one I proposed in Figure 6.8 This worksheet will help you organize the information you have gathered and assist in evaluating each waste stream and comparing various waste streams. This will also help the assessment team set priorities for waste streams at your facilities. Once you have your data in worksheets, it is customary to place the information into a relational database for sorting and analysis later on. You could also use electronic clipboards as your worksheet to eliminate the step of entering information onto a worksheet and then entering it again into a computer database. This, of course, is only important if you have a lot of data to analyze. A typical worksheet includes facility information, process information, and input information.

- The facility information should include the building number, shop name, and point of contact name, title, and phone number.
- Process information should include:
  - A unique ID or identifier number assigned to each process
  - Process title — the name of the process and the type of operation (batch, continuous, semi-batch, etc.)
  - Hazardous material used and waste generated — specify the particular chemical(s) or hazardous waste type(s), and the program goal(s) affected by these chemicals
  - Purpose — specify any directive technical orders or regulations and give a general statement of why the process exists
  - Description of the process
  - Output units and destination
  - Production rate — how many output units are processed per year

- Amount of material or waste used or generated per unit output — a number relating the amount of undesirable material to the production rate; this information improves the accuracy of shop estimates and is useful for generating ideas
- Total amount used or generated by the process
- Input material information
- Input units and origin — input/output units are what the process manipulates, the reason for its existence
- Origin (destination) is generally other organizational elements — this information describes the relationships between processes often needed for generating options and checking production rates

After you have completed these steps and are compiling data into a database or spreadsheet, it's a good time to start scheduling the site visits.

STEP 8: CONDUCTING THE SITE VISIT

Before you conduct the site visit you should have a pretty good detailed understanding of the process you're visiting. You'll probably have detailed information from the document reviews, the returned questionnaires, and any previous interviews you've conducted. The purpose of this visit is to verify the process data you have gathered, to fill any data gaps, and to gain additional knowledge from the process owners.

To minimize the impact of your site visit, visits should be well planned in advance. I recommend that you prepare a detailed agenda for the visit that outlines those areas you need assistance with. This agenda should be forwarded 5 to 7 working days prior to your visit. If you've completed an analysis, such as the summary worksheet shows in Figure 6.3, I suggest that you provide a copy with the agenda. When you schedule your visit, try to schedule it to coincide with the operation you are interested in observing or understanding, so you get a first hand knowledge of the actual process. If it is a continuous process you may want to observe the operation at various times of the work day to include all shifts. This is especially important if the operation is highly dependent on human involvement. I would recommend photographing or videotaping the operations. This will be very beneficial as you get in the opportunity assessment stage.

As you interview workers and operators you need to solicit their opinions on how much waste is generated and how to reduce that waste. I'd also try to assess how the operation you're assessing is linked to other operations in the facility and evaluate the level of coordination with other shops and departments as it pertains to environmental or waste generation issues. You may also want to consider assessing administrative controls such as accounting procedures, material purchasing procedures, and waste collection procedures.

I recommend that you observe the "housekeeping" aspect of the operations you're assessing as well. Look for signs of spills, leaks, odors, and fumes, etc. Pay attention to the overall cleanliness of the shop. Also, assess how much hazardous material is disposed of due to poor inventory management, such as expired shelf life, damage containers, inadequate material batching or mixing, etc.

Several examples of questions asked during baseline surveys are listed below. These questions may be asked continually to ensure that the most accurate information is gathered for future years. These questions can be answered during the document review and verified during shop visits.

- What is the composition of waste streams and how much waste is generated from each?
- Which wastes are classified as hazardous? Do any of the wastes or emissions fall under environmental regulations?
- What makes them hazardous? Are they included in EPA's list of target toxic chemicals?
- What are the costs of waste management and disposal for each waste stream?
- What are the environmental, safety, and occupational health hazards and liabilities?
- What amount of hazardous waste is expired shelf-life hazardous material items?
- What amount of hazardous materials is thrown away unused?
- Who is the "belly button" for each process that contributes to hazardous waste?
- From which process does the waste generate? Who owns each process?
- What are the amounts and types of municipal solid waste and hazardous materials that are presently recycled? What others could be recycled?
- What raw materials or other input materials are found in each waste stream?
- What quantity is lost as fugitive emissions (this is needed if your concern is with air quality issues).

Although the purpose of the site visit during this phase of the P2 cycle is to verify your information to date and to fill any data gaps you may have, do not limit yourself. Keep in mind that you will be trying to determine opportunities for reductions during the next phase of the process. If reduction opportunities are obvious, keep records of these. Also, solicit the inputs of the process owners, workers, and operators. These are the real experts who have the process knowledge to make substantive change to the waste generation rates.

STEP 9: ANALYZING AND PRESENTING THE DATA

Now that you have all of these data from your document reviews and from your site visits, what do you do with it? Depending on the size and complexity of the processes, operations, and procedures you may either use a worksheet or a database to collect, organize, and display your data. If you input your data into a database you should have identified a unique process identifier for each process. This will allow you to sort your data by various type. Another way to sort the data is by process codes. You could identify process codes based on the type of process that generates waste (i.e., facilities painting, vehicle painting, aircraft painting). Then you could sort the data based on those process codes.

Suggested Baseline Areas. The number of processes analyzed during the baseline assessment will determine how you present the data. I recommend that you plot a trend chart or run chart as shown in Figure 6.1 for each waste stream you analyze. Also, sort the data by process codes or at least by waste stream type. The actual waste stream you analyze most thoroughly depends on the goal, objective, or target selected from the EMS during the vulnerability phase. Some suggested trend charts include:

- EPA 17 industrial toxic pollutants (amount used)
- Hazardous wastes (amount disposed)
- Specific hazardous waste streams sorted by type of waste
- Municipal solid waste (amount disposed)
- Volatile air emissions (amount emitted)
- Ozone-depleting chemicals (amount used or purchased)
- Purchase of recycled materials: (amount used or purchased)

CLOSING THOUGHTS

The baseline assessment is a critical step taken early in your journey to move beyond environmental compliance. It is a step you must take to fully understand the magnitude of waste generated from each of your waste streams. You then use this information to determine which waste stream to further study for future P2 reduction efforts. The baseline also becomes critical for measuring success of your initiatives.

REFERENCES AND NOTES

1. In addition to my experience conducting baseline assessments I also used two documents as I developed this chapter. The documents are: a. Pingenot, J. and Voorhis, J., *The Pollution Prevention Assessment Manual for Texas Businesses,* prepared for the Texas Water Commission, 1995; b. Pollution Prevention for Federal Facilities, U.S. EPA Publication Number 300 B-94-007, 1994.

7 The Opportunity Assessment

The Opportunity Assessment (OA) is the heart of P2, but before I explain what it is, let's review what you've done so far. First, your organization developed a policy which sets the overall direction for your program. Then you developed goals and objectives which will allow you to meet your policy. Along the way you determined which waste streams you wanted to analyze or eliminate via your vulnerability assessment. Next you analyzed those waste streams, or waste-generating activities, to determine the exact amount and type of waste those particular processes generate. The OA is the next step. It's a sequential method in which you first fully understand a waste-generating process and then identify ways to reduce or eliminate the waste. Next you screen all of the proposed waste reduction alternatives to select the most feasible. Finally, you make recommendations to senior management on how to eliminate the waste. As you can imagine, the OA involves team building, data gathering, and data analysis.

As was the case during the baseline survey phase, you and your team will start analyzing existing written materials and site evaluations. However, you will delve much more deeply into each production process you have decided to further analyze.[1] You'll be interviewing workers and compiling data that you may not have collected before. During the OA, the team may identify some options that can be implemented quickly and with little cost or risk. It is likely, however, that many options will be more complex and will require in-depth analysis and some process reengineering and or benchmarking.

As I mentioned above, the OA is actually a series of sequential steps. These steps are shown in Figure 7.1 and include:

- Selecting and organizing the team
- Screening all waste streams to decide which waste streams to evaluate first
- Understanding the process by flow charting it
- Conducting site visits and detailed assessment
- Generating potential process changes
- Evaluating and screening the potential changes
- Selecting the process change and gaining approval

Each of the steps is discussed in detail below and is provided as a general guideline or a road map to follow to help you stay focused. You may have already completed some of the steps when you conducted the baseline and vulnerability assessment. If so, simply ignore them and go on to the next step. Also, as I discussed in the previous chapter on baseline assessments, it is not always clear as to when you move from the baseline phase into the OA phase. I've laid out these sequential

FIGURE 7.1 The opportunity assessment steps.

steps for both the baseline and the OA to keep you focused if you have numerous teams looking at numerous processes, perhaps even at multiple facilities or locations. If you have just one team looking at a couple of processes at one location, then painstakingly following the sequential steps is less critical to staying on track or focused, but understanding how they fit together is still necessary.

OA STEP 1. SELECTING AND ORGANIZING THE TEAM

Your success with implementing P2 initiatives is directly associated with the level of effort you put into the OA. Remember the old saying, "poor planning produces poor performance"? Well, it's very applicable, because the OA yields the plan for implementation. One of the critical steps is selecting the team. You're going to have to identify teammates who are analytical enough to study the existing process so you will understand it in great detail. But they also need to be creative enough to investigate alternative processes or procedures in order to find the best product substitute or perhaps even create a new manufacturing process. The OA team may be the very same team you used to do the baseline assessment or it could be an entirely new team. If it's a new team, it should include at least one member of the baseline survey team. Although this person does not have to lead the OA phase, having at least one person from the baseline team provides a more seamless transition into this phase. Unless your organization is very small you'll need some additional staff to comprise one or more opportunity assessment teams. If you have multiple teams, the focus of each assessment team will be relatively specific; one team may focus on hazardous waste, while another may focus on air emissions. Another way to look at it is to have each team focus on a different process, such as one looking at the photoplating shop while the other studies the photographic shop process.

Most teams have three to six people; more than that become unmanageable. Please remember that technical specialists can be consulted if needed. As I mentioned above, one member of the baseline survey team will be included on each OA team to facilitate communication and provide a brief history of efforts completed to date. The rest of the team should include people who are directly responsible for, and therefore possess the most knowledge of, the waste streams and the areas of the facility you're considering. With that in mind, the first team member you select should be the "process owner," regardless of his/her stature within the organization. The closer to the process the better! In fact, I have observed many teams comprised of engineers and operations managers being led by a process owner who's an hourly employee. Rounding out your multidisciplinary team to the extent practical, you should consider engineers, supervisors, and other production workers as well as finance and accounting, purchasing, and administration staff when selecting the team members.

Please do not just consider a person's technical expertise or field of expertise when picking a teammate; also consider a candidate's other skills, such as ability to work on a team, interest in, and commitment to P2, and the capacity for creative thinking and solving long term problems.

Just like during the baseline stage, before you select a team, you need to decide on the level of effort you plan on taking so you know which skills you'll need on your team. Ask yourself the following questions:

- Are you going to analyze all of your waste streams?
- Are you analyzing just the largest waste-generating waste streams?
- Are you analyzing the waste streams which are most staff-hour consuming?

These decisions must be made by your environmental steering group after being presented with the results from the baseline survey stage. The decision must be tied into the EMS goal and objectives as discussed in Chapters 5 and 6. Knowing your organization's goals and objectives is important, because they drive your team's behavior. Is your organizational focus on reducing operating cost, or is it to protect the natural environment? Perhaps the focus is on protecting human health? Each of these organizational goals are fine goals, but they drive different activities. For example, if your goal is to reduce operating costs, your organization may take a "most costly to dispose" approach to reducing hazardous waste. If the organization's goal is to reduce any risk to human health, the focus may be to eliminate the purchase and use of all toxic materials. Also, the skills needed by your team members may be different depending on which of the goals you're trying to achieve. I highly recommend that your organization should sort out its objectives and goals prior to pressing further ahead with the OAs. Without this detailed effective planning, you are doomed to poor performance.

OA STEP 2. SCREEN AND RANK ORDER OF THE WASTE STREAMS TO DETERMINE WHICH TO EVALUATE FURTHER

As I have mentioned numerous times, great detailed analysis of every waste stream may not be practical for your organization. So focus on the worst first. As discussed above and during the baseline assessment phase, each organization defines "worst" differently. The key points are to ensure maximum linkage to your EMS and your

WASTE STREAM ID:			
PRIORITY RATING CRITERIA	RELATIVE WEIGHT (W)	RATING (R)	R x W
1. Cost of Disposal			
2. Environmental Regulations			
3. Raw Material Costs			
4. Threat to Workers			
5. Threat to the Environment			
Sum of Priority Ratings. SUM (R x W)			

FIGURE 7.2   A correlation matrix for screening waste streams.

objectives. As you screen your options, you may realize that some have no risk or cost attached to them and should be implemented immediately without further analysis. In fact, housekeeping initiatives usually fall into this category. Other options may prove to be impractical and should be dropped from consideration. The remaining options usually get further study through feasibility analysis. The screening does not require detailed and costly study. Screening procedures can range from applying an informal review to formal decision-making tools. Many screening tools have been discussed in Chapter 5 on vulnerability assessments. One tool to use to do this is a correlation matrix, such as shown in Figure 7.2. With this tool you simply list all of your waste streams on the left (rows) and rank all areas, such as waste streams, toxic usage, processes, and operations (columns), and select those areas targeted for further assessment based on the selected criteria. Weigh each column with a 3, 5, 7, 9, then sum the total — the highest score gives an idea of where to focus. Some of the common columns include:

- Quantity of waste generated
- Type and hazardous nature of the waste
- Treatment and/or disposal costs
- Occupational health and safety considerations
- Potential for implementation
- Quantity of toxic substance used
- Regulatory compliance (current and expected future)
- Position on pollution prevention hierarchy
- Reducing energy usage
- Available budget/ROI (payback) time
- Opportunity for economic savings
- Opportunity to perform process improvement/optimization
- Opportunity for quality improvement
- Ease of implementation

I would use this tool and perhaps one other, such as those presented in Chapter 5 and Chapter 6, to help prioritize your processes. Once you have prioritized your processes, it's a good idea to inform the process owners and the environmental steering group of your prioritized list as well as to provide a roadmap for the remaining actions. You'll need their buy-in, so getting them involved early will ensure that they support your recommendations later on. This will be discussed in greater detail in Chapter 8.

OA STEP 3. UNDERSTANDING THE PROCESS

Now that you've prioritized the processes and gained senior management support and approval, it's time to really understand the process. It's essential that you have a detailed understanding of the process before you start recommending changes to it. The flow chart or process map, along with the input–output and material-balance diagrams are great tools to use to conceptually understand a process. Don't go overboard with the information presented on these tools. The detail should be proportional to the complexity of the process you're trying to understand.

As I'm sure you're aware, flow charts are a visual means of organizing data on energy and material flows and on the composition of the inputs and outputs of a process or other activity. A material balance, in its simplest form, is a check to make sure that what goes into a process (e.g., total mass of all raw materials, water, etc.) leaves the process (e.g., total mass, wastes, by-products, etc.). Material balances are particularly useful for quantifying fugitive emissions on other very detailed analyses. In general, the level of detail of these diagrams and material balances will vary depending on the processes, operations, and your facility and its resources. As a general rule of thumb, sufficient information should be provided to enable you to understand the overall flow of materials.

THE INPUT–OUTPUT DIAGRAM

The first step in understanding a process flow is to develop a simple input–output diagram as shown in Figure 7.3. First you'll want to identify all of the inputs that could be identified, including the containers, waste that is reused, and process water. Next, I would identify all of the outputs of the unit process. This includes the products, by-products, and wastes to be recycled and disposed of.

It's important that the assessment team looks at all possible avenues of waste generation, not simply just the process itself. The "input" portion is where materials, usually hazardous or toxic materials, are stored. Obvious waste includes leaks and spills from containers and contamination from contact with other chemicals. Other possible wastes can come from expired-shelf-life or material spoilage. Do not discount this "input" step in any process. In many cases, the waste generated during this step is a significant portion of a total waste stream, and can be eliminated completely with a little more attention to "housekeeping" and warehousing discipline. Figure 7.4 is provided to assist your gathering information on the "input" step.

The next part of the input–output diagram you should complete is the output stage. This includes the product and the waste generated. Most of this information

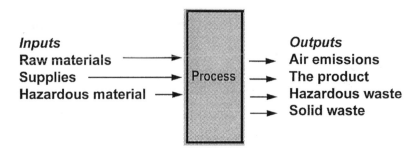

FIGURE 7.3 The input–output diagram.

was gathered during the baseline survey stage. You should also consider describing where the storing and shipping of the product takes place. Waste can still be generated during this phase, especially due to improper handling of products and poor inventory control. The shipping and transportation process also generates waste. You can get this information from bar code records, shipping records, and billings for such things as solid waste disposal, municipal water and wastewater treatment, hazardous waste manifests, MSDS, permits, Tier I reports, etc. Most of this information you probably gathered during the baseline survey stage. After you have completed this step, you should really evaluate the actual process.

THE PROCESS FLOW CHART

Once you have completed an input–output model, it is time to lay out all of the process in greater detail (Figure 7.5). This process flow chart is simply an expansion of the "process" block of the input–output model. This step will probably require more time and effort than your team expended in the "input" part. Waste, by its very nature, is caused by inefficiencies in the process. This may be caused by equipment, personnel, or both. Of course, all manufacturing organizations want to have their equipment operating in the most efficient manner possible in order to increase effectiveness, but much equipment and many processes are extremely variable. By identifying the optimum operational level and maintaining statistical process control, the waste can be significantly reduced. To identify the inefficiencies, the entire process will probably require its own detailed flowcharting. Sometimes it is easier to have one or two people complete this step instead of the entire OA team. Once the process is flow-charted it can then be reviewed and approved by the entire OA team.

MASS BALANCE DIAGRAMS

If you need a very detailed understanding of a process, it's now time to complete the mass balance diagram. Like the flow chart, the mass balance approach helps you to really understand a process. The difference is that the mass balance approach is more quantitative. With this approach, the weight of all materials entering the process must equal the weight of all materials leaving the process. When the final product which is produced weights less than the sum of the materials entering the process,

ATTRIBUTE ■	STREAM #	STREAM #	STREAM #
1. Name /ID			
2. Source / Supplier			
3. Component/Attribute of Concern			
4. Annual Consumption Rate Overall Component of concern Purchase Price Overall annual			
5. Delivery Mode			
6. Shipping container size and type			
7. Storage mode			
8. Transfer Mode			
9. Empty Container Disposal Management			
10. Shelf Life			
11. Acceptable Substitutes?			
12. Alternate Supplier?			

FIGURE 7.4 Input material summary table.

the difference is the waste generated in the process. Of course the waste may be in a number of media (air, water, solid), so you'll want to locate each of these. In its simplest form a material balance is developed according to the mass conservation principle:

$$\text{Mass in} = \text{Mass out} - \text{Generation} + \text{Consumption} + \text{Accumulation}$$

If you're going to use the mass balance approach, I would first develop the input–output diagram, then develop the process flow chart prior to developing this very detailed step. Once you are comfortable with these two diagrams, you can

- List the operations at this shop/building:

- What are the major products?:

- Provide a rough estimate of production volume per month (e.g., the number of circuit boards produced).

- What are the major waste streams? What media do they affect? How much waste is generated in each?

- Has the shop/building already implemented any waste reduction or recycling techniques? If yes, describe the technique results to date.

- How many people work in this shop/building? How many hours is it operational?

- The following sources contain information that may help you determine the type and amount of waste generated at your shop:
  - Work Flow Diagram
  - Hazardous Waste Manifests
  - Emission Inventories
  - Annual/Biennial Reports
  - Permit/Permit Applications
  - Material Safety Data Sheets
  - Chemical Materials Inventory
  - Permitting Details/Conditions
  - Environmental Audit Reports

FIGURE 7.5   Other information and questions.

continue using the information collected during the baseline assessment to develop a block flow diagram. You can continue to expand on this by adding a simple narrative, and a simple material balance for each process step, operation, procedure, or maintenance practice you plan on studying. In this manner, the input–output diagram becomes a mass balance diagram for the entire process. In this way, material inputs can be tracked and waste losses discovered. As you go through understanding the process, you will identify ways in which it can be improved or changed and identify these processes with the most potential for prevention applications.

Once you have completed all of the diagrams, the entire OA team, as well as the process owners, should review the flow diagrams, narratives, and material balances to identify data gaps, conflicts, and areas where more information is needed. You can obtain the missing information or make estimates as necessary, and incorporate into the diagrams, narratives, or material balances. If you have many assessment team analyzing processes, it would be a good idea for all OA teams to get together and prioritize the waste streams, toxics used, operations, and processes, to ensure that you have a consistent approach to this step of the OA phase. Ideally, all waste streams, toxics usage, operations, and processes should be evaluated in detail to identify pollution prevention opportunities, but this is not always practical.

OA STEP 4. SITE VISITS AND DETAILED ASSESSMENT OF THE RANKED PROCESSES

Overall, the detailed assessments and site visits are intended to provide more comprehensive information and data on the targeted waste streams. This information is crucial to identifying P2 opportunities, because it will identify where the potential for source reduction or recycling exists, and provides a general indication of the feasibility of the various P2 initiatives.

Now that you have selected the "worst-first" waste streams and have selected processes, operations, waste streams, and toxics usage for further evaluation, it's time for your team to visit the shops which create the waste so you can get intimately familiar with all of the processes. This is achieved by performing a thorough site review and interviewing numerous personnel who are involved with the processes.

The site or shop visits can also be used for validation of the existing information you have already gathered and analyzed. Since at least one of your team members is one of the process owners, the information gathered to date should be very accurate. You could present your process flow chart to the shop members you are visiting to ensure that it accurately describes the process. After they concur with the "as-is" flow chart, you can then use the site review to verify the actual operation or processes, to answer any questions that may have arisen during earlier visits or discussions, and to obtain more detailed information. Quite often you will find that the people closest to the process are the individuals who have the best ideas on how to eliminate the waste streams. Don't discount their input. They not only know where the inefficiencies are; they also know how to improve or eliminate them!

PLANNING SITE VISITS

Site visits need to be well planned to reduce the impact on the operation being assessed. While multiple site visits to the same shop may be necessary, effective planning should keep this to a minimum. Although the steps I discuss below may have already have been done in Steps 2 or 3, you should ensure that they are completed before the site visit. I also recommend completing the steps below, as well as the worksheets in Figures 7.5, 7.6, and 7.7, to help you plan your site visit.

- Review existing documentation such as data on waste generation, equipment operating manuals, Tier III reports, permits, and shipping and reviewing records.
- Decide on data collection format to make sure the data collection is rigorous and in compatible format with further analysis you may be conducting.
- Send out the questionnaires to the shops being visited prior to your site visit. Ideally, you will want to get the questions returned and conduct some analysis prior to the site visit.
- Obtain a camera and a video camera to get pictures of the process.
- Schedule the site visit at least 2 weeks prior to the visit.
- Prepare an agenda and send it out at least 1 week early.

Questions	Waste Stream #	Waste Stream #	Waste Stream #	Waste Stream #
Waste Stream ID / Name or Process Unit				
Brief Description of how waste is generated				
Component or Property o Concern				
Annual Generation Rate (Units:)				
Waste Stream Type: (A) air emission (WW) waste water (SW) solid waste (HW) haz waste				
Occurrence (R) regularly (NR) non-recurrent				
Monthly generation rate				
Treatment/Disposal method (R) recycled (L) landfill (S) sewage treatment plant (I) incinerator (O) other				
Disposal Cost $ per: Annually				

FIGURE 7.6   Waste stream assessment form.

- When conducting the site visit, observe the actual operation by all operators on all shifts, since there could be great variation in the operation between shifts and personnel.
- Interview worker and supervisors. If possible provide them a copy of the questions you plan to ask prior to the interview.
- Follow the entire process from beginning to end. The assessment should cover all of the process flows and operations including the following steps:
  - Shipping and receiving areas
  - Raw material storage area
  - Unit processes and product/by-product areas
  - Known waste generation points

Questions	Y	N
GENERAL		
• Does the shop have written procedures for shop inspection or maintenance?		
• Does the shop show signs of poor housekeeping (cluttered walkways, uncovered drums, etc.)?		
• Does the aisle space appear adequate for equipment, bulky items?		
• Is the drum storage area covered (e.g., all containers are protected from the weather)?		
• Are all drums raised off the ground (e.g., on pallets)?		
• Does the shop have equipment to prevent releases caused by over-filling of storage tanks (e.g., high level shutdown/alarms, secondary containment)?		
• Are there noticeable spills, leaking containers, or leaking valves, pumps, hose fittings?		
• Are containers labeled as to their contents and hazards?		
• Is there smoke, dust or fumes indicating material losses?		
• Does the shop have a training program for raw materials handling spill prevention, proper storage techniques and waste handling procedures?		
• If this is a training facility, are students trained to be environmentally aware?		
INVENTORY CONTROL		
• Does the shop have a single point of contact for ordering and distributing new supplies?		
• Does the shop have written procedures regarding inventory control?		
• Is inventory used in first-in, first-out order?		
• Do you dispose of products due to expired shelf-life?		
• Are empty containers returned to the supplier, or re-used on-site?		
• Does the shop buy products in bulk?		
Estimate the volume of chemicals maintained on-site:		
• If figures are unavailable, provide qualitative assessment: Low, Medium, High (Low = one supply cabinet Medium = 1 to 3-55 gallon drums High = stock room with more than 50 products with > 5-55 gallon drums)		
• Check the inventory quickly - are there any toxic substances?		

FIGURE 7.7   Management practices questionnaire.

Incorporate any new information or data into the flow diagrams, narratives, and/or materials balances for each of the targeted areas. Based on the information obtained from the baseline and detailed assessment, develop a description for each that will be the focus of your pollution prevention efforts.

WASTE STREAM ID:			
PRIORITY RATING CRITERIA	RELATIVE WEIGHT (W)	RATING (R)	R x W
1. Achieve P2 Goals			
2. Technology Exists			
3. Timely			
4. Implementation Costs			
5. Reduce Threat to the Environment			
6. Training Required?			
7. Regulatory Compliance			
8. Treatment Disposal Costs			
9. Potential Liability			
10. Waste Quantity Generated			
11. Waste Hazard			
12. Safety Hazard			
13. Potential to Remove Bottleneck			
14. Others			
Sum of Priority Ratings. SUM (R x W)			

FIGURE 7.8   A screening matrix for selecting an option.

Site Visit Questions

Typical questions asked during site visits are provided in Chapter 5 and Chapter 10. Review and modify the questionnaires to best fit your needs and forward them to the process owners prior to the site visit. Figures 7.6, 7.7, and 7.8 are provided to assist you. In general, the questions typically focus around these basic areas.

- How much and what is the composition of the waste stream and emissions generated in the process?
- From which processes or treatments do these waste streams and emissions generate?
- Which waste and emissions fall under environmental regulations?
- What raw materials on input materials are used in the process? Which of these are hazardous?
- How much of each specific raw material is found in each waste stream?
- How much waste is lost in terms of volatile emissions?
- What is the efficiency rate of the various portions of the process?
- Are any unnecessary waste materials or emissions produced by mixing materials which could otherwise be reused?
- What process controls are in place to increase production efficiencies?
- Do the process owners have any ideas on how to eliminate or reduce the waste stream?

OA STEP 5. GENERATING POTENTIAL
PROCESS CHANGES

After you have completed the site visits and feel you fully understand the processes in question, it's necessary to develop improvement ideas and evaluate the targeted areas in terms of the source reduction or recycling opportunity available.

This is the creative phase of the P2 analysis during which the OA team proposes and then screens opportunities for each area. Brainstorming sessions are very useful during this phase, allowing for the proposal of a wide variety of pollution prevention opportunities. At a minimum, however, your team should consider the following opportunities for each targeted area:

- Input changes or material substitutions
- Operational improvements or improved housekeeping
- Production process changes/process optimization
- Product redesign or reformulating
- Closed loop recycling systems
- Direct reuse or reclamation systems
- Best management practices

At the present, there are numerous sources of information available to help you identify pollution prevention opportunities, both general and industry specific. Chapter 12 is a reference chapter which I've provided to assist in this area. The very first source of information you solicit should be from individuals within your organization. This should include line personnel and employees, operators, supervisors, engineers, plant managers, purchasing agents, and others with first-hand knowledge of the operation. In addition, potential outside sources of information include:

- State/local environmental agencies' publications and technical assistance programs
- EPA publications, databases, and technical reference centers
- Published literature, technical reference centers
- Equipment vendors and chemicals suppliers
- Trade associations
- Benchmarking clearing houses

OA STEP 6. SCREENING THE POTENTIAL
OPPORTUNITIES

The initial set of proposed opportunities must now be screened to eliminate those perceived to be marginal, impractical, or inferior prior to conducting more detailed and time-consuming feasibility studies. This preliminary screening may consist of an informal decision made by the team leader, or a more analytical method such as the use of a weighted sum method of screening. Questions to consider during the screening include:

- Which opportunities will best achieve the organizational goal or objective you are attempting to achieve?
- What are the main benefits to be gained by implementing this opportunity?
- Does the necessary technology exist to develop this opportunity?
- Can the opportunity be implemented within a reasonable amount of time?
- What other areas will be affected, such as product quality?
- Does this opportunity have a good or proven "track record"? Is there strong evidence that it will work as required?
- Is this opportunity compatible with the existing facility operation?
- How much downtime, training, etc. will be required to implement this measure?
- How much does the proposed change cost? Does it appear to be cost effective?
- How high or low are the current disposal costs? Are these costs going to increase?
- Examine current environmental regulations and future regulations. Do these laws inhibit current processes now? How about in the future?
- Are the costs of raw materials high or are the prices likely to increase significantly in the near future?
- Does the waste generated or the material used in the process pose a threat to your workers?
- Does the waste create complex problems for processing, handling, storing, and discharging?
- How much volume is generated or released and will future capacity be a problem.

This preliminary screening should result in an intermediate set of P2 opportunities, i.e., "intermediate opportunities." The "intermediate opportunities" are those that have merit and warrant further analysis to determine if they are viable "options" for implementation. Based on the information obtained during identification and screening of your P2 opportunities, develop a description of each "intermediate opportunity" that will be further evaluated. Use the correlation matrix shown in Figure 7.8 or those presented in the chapter on vulnerability assessment to assist with this step.

OA STEP 7. THE FEASIBILITY ANALYSIS

From the screening matrix you now have a good idea of which ideas to further evaluate. The P2 "intermediate opportunities" resulting from the brainstorming and preliminary screening phase of this analysis will be further evaluated to determine if they are feasible "options." In general, you should complete some level of economic, environmental, and technical feasibility studies for each intermediate opportunity.

TECHNICAL EVALUATION

The technical evaluation examines each opportunity to consider whether or not it could be incorporated into current operations. Your employees who are most familiar with the process and equipment can contribute valuable ideas. Quite often simply soliciting employees' ideas will identify easily implemented, common sense solutions. If you consider process changes, they may need to be pilot tested or at least researched. Consider these factors when you're completing a technical feasibility study:

- Will it work in this application?
- How has it worked in similar applications?
- Is space available? Are utilities available or must new utilities be installed?
- Is the new equipment or procedure compatible with the facility's operating procedures, work flow, and production rates?
- What is the downtime associated with installation or implementation of the opportunity?
- Will the product quality be maintained or improved?
- Is special expertise required to operate or maintain the new system?
- Does the vendor provide acceptable service?
- Are any new safety hazards created?
- How soon can the system or process change be implemented?
- Will the installation procedures stop production? If so, for how long?
- Will training be required?
- Does the new system alleviate human health and environmental problems?
- Are there any regulatory barriers? Are permits required?
- How have you involved maintenance and engineering in the technical evaluation?
- Does the new process require a change in labor? Is the Union involved?

ECONOMIC EVALUATION

Once you have established that an option is technically feasible and it will in fact reduce your waste generation, then your next step is to determine if the change is economically feasible. This is where the economic evaluation comes into play. If, of course, the process change is extremely simple, this step may not be required. It may be obvious that the process change will save you money, but for more complicated improvements you may need to do a more detailed analysis, such as determining the payback period and return on investment. You will need to gather cost data for the more complicated opportunities. Cost becomes important when you're selecting pollution prevention actions from a list of opportunities, or when project justification must include economic cost/benefit data. First, identify all costs associated with each hazardous material purchased or waste generated. These should include not only purchase and disposal costs, but also costs of special process equipment, permit applications, personal protective equipment, training, etc. directly attributable to the material. Establish a total cost per unit weight for each chemical and each type of waste. For a data integrity check, ensure:

- Calculated process usage and generation rates total equal the amounts reflected in purchasing and disposal contract records
- Material quantities issued for each organization minus unused quantities equals production rates times usage per unit output

- Waste quantities generated by each organization equal production rates times generation per unit output
- Sequential processes have the same production rates

Consider these factors when you're completing an economic feasibility study:

- Reduced hazardous waste costs (disposal, management, etc.)
- Raw material cost savings
- Insurance and liability savings
- Increased costs (or savings) associated with product/service quality
- Decreased (or increased) utilities costs
- Decreased (or increased) operating and maintenance
- Costs due to cycle time changes
- Decreased (or increased) overhead costs
- Capital costs
- Manpower costs

I recommend that you evaluate capital costs, incremental operating costs, and profitability and bayback. You will want to involve somebody from your accounting office to help you complete these analyses, but I have provided some simplistic outlines to assist you in getting started.

- *Capital Costs.* Capital costs (Figure 7.9) are the initial expense of getting the new process built, installed, and operational. They include both the fixed capital costs for designing, purchasing, and installing equipment, as well as costs for working capital, permitting, training, start-up, and financing. Complete the worksheet for each opportunity you are evaluating.
- *Incremental Operating Costs.* Incremental operating costs and revenue are the differences between the estimated costs of your P2 option and the actual operating costs of the existing system before the option is installed (Figure 7.10).
- *Profitability and Payback.* As you know, payment period is the amount of time it will take for you to recover your initial outlay. There are various ways to calculate this but I have proposed a very simplistic approach. For a more detailed analysis I recommend you talk with your accountants. Usually, projects with a payback period of three to four years are reasonably acceptable. Longer than that may need further analysis. The formula is:

Payback period in years = Total Capital Investment/Annual Net Operating Cost Savings

- *Internal Rate of Return.* One method you can use is the Internal Rate of Return (IRR) for ranking your capital investments against other projects competing for funds. Although the IRR is referred to as the payback period

ITEM	COST
Direct Capital Costs	
Site Development	
Process Equipment	
Materials	
Utilities and Services	
Construction and Installation	
Indirect Capital Costs	
Engineering, Design, Procurement	
Permitting	
Contractor's Fees	
Start-up	
Training	
Contingency	
Interest Accrued during construction	
Total Fixed Capital Costs	
Working Capital	
Raw Material Inventory	
Finished Product Inventory	
Materials and Supplies	
Total Working Capital	
Total Capital Investment	
Salvage Value	

FIGURE 7.9   Capital costs table.

as a method for determining economic viability of your project, its calculation is usually very involved. I will not go into detail, as this method is beyond the scope of this text.

Environmental Impacts of the P2 Option

When evaluating a potential opportunity, you should also evaluate the environmental impacts of your decision such as:

- What is the effect of this opportunity on the number and toxicity of waste streams?
- What is the risk of transfer of pollutants to other media?
- What is the environmental impact of alternative input materials?
- What is the associated energy consumption?
- Are any new environmental hazards created?

Operating Cost/Revenue	$ per Year
Operating Costs	
Decrease for Disposal	
Decrease in Raw Materials	
Decrease in Utilities	
Decrease in Labor	
Decrease in Supplies	
Decrease in Insurance	
Total Decrease in Operating Costs	
Revenue	
Incremental Revenue from Decreased or Increased Production	
Incremental Revenue from Marketable By-Products	
Incremental Revenue	
Net Operating Cost Savings (loss)	

FIGURE 7.10   Incremental operating costs and revenue worksheet.

SOURCE REDUCTION CONSIDERATIONS

Consider these factors when considering a source reduction option.

- Inventory control and material and waste-tracking systems
- Labeling all containers, using bar-code systems to monitor
- Improve scheduling of batches to reduce the need for cleaning
- Record-keeping: documentation of process procedures, control parameters
- Preventive maintenance — regular inspections and corrective maintenance
- Spill and leak protection
- Material usage, handling, and storage
- Standardized materials (paints, solvents, cleaners)
- Spacing containers, labeling containers, separating hazardous material
- Employee education: training in the proper operating procedures and regular emergency response drills

OA STEP 8: SELECTING THE OPTION

Based on the steps you've completed, you have determined whether or not the pollution prevention opportunities should be implemented. These are the intermediate opportunities that are in fact feasible and could be implemented at your plant. Please note that although an opportunity may be feasible, it does not necessarily mean that it must be implemented. For example, there may be more than one pollution prevention option available for a particular waste stream. All options may be feasible, but only one will actually be implemented. Also, it is probably not possible for you to implement all feasible options at once. Choose on option and press ahead; Figure 7.11 is provided to help you summarize. Chapter 8 explains the next steps.

CLOSING THOUGHTS

The purpose of the OA phase is to develop a set of prevention options for each waste stream and then to identify the most attractive options that deserve a more detailed analysis. The OA is a very systematic approach to first understanding the existing or "as-is" process, then identifying opportunities to reduce the waste generated from the process. First, you review the EMS to determine which waste streams you attack first. This depends on the strategic direction set forth by the policy and the EMS. Don't try to attack all of your waste streams. Just attack one or two at a time. Next, get leadership approval of your proposed action plan and select the OA team. Make sure you have the process owners involved! Then you begin to understand the existing process which generates the waste. The process owner must flowchart the existing process. Then the OA team can draw a simple process mass balance diagram to understand what happens to the materials used in the process. You may ask the consumer of the product you're producing what their actual requirements are. If you're cleaning a part, how clean does it need to be? Once you understand the existing process then you can begin identifying possible opportunities. You can brainstorm possible alternative methods. Involve the shop personnel at this stage. You may also want to conduct a benchmarking study or literature search to determine if there is a better way to do business. From there you develop a list of proposed options and screen the options for technical, environmental, economic, and manpower constraints. Some TQM tools you can use in this step include the flow chart to understand the process; the Gannt chart for the proposed action plans; and benchmarking. These and other tools will make your analysis move along without stagnation.

Opportunity Title: _____

_____

Describe the Opportunity: _____

_____

_____

Waste Stream(s) Affected: _____

_____

Input Material(s) Affected: _____

_____

Product(s) Affected: _____

_____

Indicate Type of Opportunity: (the opportunity may be a combination of source reduction, recycling, and treatment) _____

_____

Source Reduction (indicate type)
_____ Equipment Change
_____ Personnel/Procedure Change
_____ Material Change

Recycling/Reuse (indicate type)
_____ Onsite
_____ Offsite

Treatment/Disposal.
After source reduction and/or recycling have been implemented, residual wastes may still require treatment and disposal (e.g., still bottoms or spent filters from an on-site solvent recycler).

Treatment Required.
_____ Biological   _____ Incineration   _____ pH adjustment
_____ Precipitation   _____ Solidification
_____ Other

FIGURE 7.11   The opportunity description.

Opportunity Assessment (Cont.)

Explain why the opportunity appears to be technically feasible: _____
_____
_____
_____

Estimate the one-time cost to implement the opportunity, taking into account equipment
costs, labor, etc.: _____
_____
_____
_____

Estimate the annual recurring costs to implement the opportunity, taking into account
materials, manpower, utilities, etc.: _____
_____
_____
_____

Estimate how much money implementation will save the facility annually. _____
_____
_____
_____
_____

Would the opportunity be difficult to implement?   Y   N
    Staff training required?   Y   N
    Staff/management resistance expected?   Y   N
    Cost to implement is high?   Y   N
    Labor is required to install and maintain new equipment? Y   N
    Other factors (please add).   Y   N

FIGURE 7.11 (continued)

REFERENCES AND NOTES

1. There are numerous federal and state publications available on conducting polution
    prevention studies. Two that I particularly like and used as I developed this chapter
    are: a. Pingenot, J. and Voorhis, J., *The Pollution Prevention Assement Manual for
    Texas Businesses*, prepared for the Texas Water Commission, 1995; b. *Pollution
    Prevention for Federal Facilities*, U.S. EPA Publication Number 300 B-94-007, 1994.

8 Implementing Your Opportunities

Now that you've completed the studies, it's time to get into the execution phase of the implementation. If you recall the P2 cycle I presented in Chapter 5, shown again in Figure 8.1, you can see that the next steps include:

- Step 5. Developing the implementation plan
- Step 6. Implementing the plan
- Step 7. Measuring the results
- Step 8. Standardizing the solution

CIP STEP 5. THE IMPLEMENTATION PLAN

The implementation plan essentially involves detailing the specific actions required to implement the OA recommendations. Typically it takes the form of developing a proposal for the option you've selected and recommending that option to senior management for approval and commitment of resources. The proposal will probably be in terms of a written report, summarized with an oral presentation to present the results of the studies, the technical feasibility analysis, and the economic feasibility analysis. Remember, one of the key parts of implementing an EMS is for senior management to allocate the appropriate resources. It's up to you to articulate what resources are required.

The proposal should convince senior management to fund and implement your P2 option. It should review all phases of your assessment and clearly describe the amount of data gathering and analysis you have done to get to where you are. I recommend that you create a storyboard to outline the steps. The storyboard will demonstrate the level of effort you have exerted to get to the point you are at. By covering these steps before recommending actions, you will create confidence and credibility in your proposal.

The proposal should also detail the options that your team has determined are most feasible and recommend the schedule for implementation. As you develop your proposal, keep in mind that you should evaluate each alternative under different scenarios such as the most optimistic and pessimistic cases. When appropriate, sensitivity analysis should be applied to determine the profitability of various alternatives.

The level of detail in each proposal varies with the type and complexity of the project. For example, in the municipal solid waste (MSW) area, your suggested proposal may be to purchase and install two cardboard-baling machines. Examples of actions that should be considered when executing this project might be identifying the location, who will be operating them, what training the operators need, what procedures will be followed to use the machine (e.g., asking customers to segregate

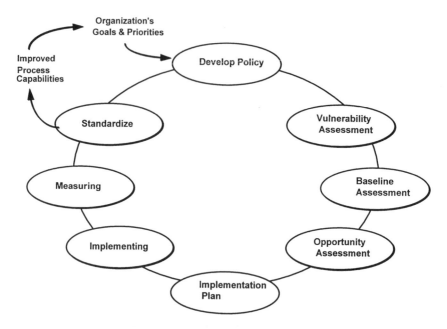

FIGURE 8.1   The P2 cycle.

cardboard waste, how bales will be transported, etc.), and special facility require-ments such as electrical power supply, concrete pads, etc.

Before you submit the proposal, ensure that you coordinate it with the affected departments to solicit their support and approval. If your option involves operational, procedural, or material changes, it will receive a better chance of being accepted by senior management if you can provide extensive information and show that it has been coordinated and approved by all affected parties. Your proposal should discuss these points:

- What will this project do? Does it save money, manpower, reduce pro-duction costs, increase public confidence, reduce environmental liabili-ties?
- Which EMS goals and objectives does the proposal support? How does the proposal link to the organizational environmental policy?
- What is the project? Discuss location, equipment, personnel, timing of implementation.
- What will the project cost? Include both the initial capital investment and the net operating costs. Discuss the overall project economics, and where the resources will come from.
- Is the project profitable?
- How does it impact the environment? What is its P2 potential?
- What is the estimated time for installation and start-up?
- What are those possible performance measures you'll use to determine success?

As you answer each of these questions, remember that your management may not have the same vision you have for prevention-related improvements. Therefore, you may need to "sell the project" by carefully describing what the project will do for the company, operating division, or plant. Remember to explain any reduction of environmental risk and any financial benefits that the project may have. Also, explain in great detail how this P2 initiative helps the organization achieve its environmental policies, goals, and objectives. If possible, show what other organizational objectives your proposed initiative achieves and discuss which stakeholders receive the benefit.

THE REPORT LAYOUT

I'm sure your company has a layout for reports and proposals. If not, here's a layout to consider.

- Executive summary
  - State the problem
  - State the proposed solution
  - State the benefits of the solution (economic, legal, other)
  - Indicate how it achieves the EMS Policy
- Introduction
  - State the problem your project will solve. Don't forget to include any of your waste assessment data needed for support as well as the regulatory situation
  - State the proposed solution. Brief introduction to the proposed solution
  - Give a statement relating the problem and solution to your company environmental policy and your EMS goals and objectives
- Environmental review
  - Chemical compound characteristics
  - Human health risks
  - Risks to the environment
- Technical review
  - Review the proposed technology. Mention successful applications
  - Identify any facility changes or modifications
  - Identify any process or material changes
  - Discuss the construction period and down time (if any)
  - Discuss how to evaluate a project's success
- Economic review
  - Review the proposed capital costs
  - Review the proposed net operating savings (costs)
  - Review the payback period and/or IRR and sensitivity analysis
- Structure and responsibilities
  - Assign roles and responsibilities for implementation with as much detail as possible
  - Review the required resources (people, funding, equipment, material) and how they will be obtained

- Training and awareness
  - Identify any special training needs and who should receive the training
- Proposed measures
  - What the measures are, how they will be presented
  - Who will collect and track the data
  - How the information will be communicated between departments
- Recommendations and conclusion
  - Restate the problem and proposed solution
  - If warranted, make recommendations such as: further assessments; contractors; and changes to policy, goals, or objectives.

THE FORMAL PRESENTATION

Don't develop your presentation until you've completed the written report. You should present what is contained in your report, so ensure that there is great consistency between the report and the presentation. Your presentation should follow the outline I presented above for the report. Before you brief the environmental steering group, I recommend that you pre-brief the process owners. Even though they have been involved throughout the entire study, it continues to build their buy-in, thus making the implementation easier later on.

FINAL DOCUMENTATION

Once you have completed your report, presented your ideas, and received approval, you are not quite finished. Take some time to document the decision-making process and the decisions made by the steering group. Carefully log, catalog, or otherwise properly store the work you have done so far. Chances are your team or another team will use it again when they implement the remaining opportunities. Additionally, your environmental compliance people probably could use the information for the never-ending regulatory reports they are responsible for completing, and your EMS manager can use the information for the EMS documentation and eventually for the EMS audit.

CIP STEP 6. IMPLEMENTING THE PLAN

Now that senior management has approved and funded your project, you can get out of the planning stage and into implementation. If your organization doesn't have a set of guidelines to follow I think these steps are helpful,[1]

- Design the system. Involve people from production, maintenance, safety, and of course, the process owners, plus any other people who may have inputs on layout, scheduling, and equipment selection.
- Prepare the construction bid package and equipment specifications (if required).
- Select the construction firm and materials.
- Install the equipment.

- Train the users.
- Start the operation slowly.
- Monitor and evaluate performance.

The first step in the implementation of the option is to record the start date and the last step is to record the completion date. Between these two steps your implementation plan should have milestone dates for each of the major tasks. A couple thoughts to consider during implementation: make sure you get guarantees from vendors for any equipment and supplies you purchase and, also, your vendor should allow you time to bench-scale-test their equipment and may even provide the training free of charge.

CIP STEP 7. MEASURING THE RESULTS

Measuring success of your P2 initiative can be a complex process. You may need to combine some of the methods I present below to meet the needs of your particular waste stream or organization. Some of the methods to measure reduction trends include measuring the:

- Quantity of waste shipped off site or treated on site
- Quantity of hazardous materials used or received
- Quantity of waste generated
- Change in the amount of toxic constituents
- Change in the overall material toxicity

The easiest way to evaluate the performance of your waste reduction project is to:

- Measure if it meets its expected economic benefits; and
- Measure waste reduction by recording the quantities of waste generated before and after the process is implemented.

Measuring the economic benefits can be made by comparing your actual realized savings and internal rate of return to those projected in your economic analysis as discussed in Chapter 7. The measurement of waste reduction is more involved. You will need to select a quantity (weight, volume, toxicity) of waste to measure, measure that quantity, and normalize the data as necessary to correct for factors not related to the P2 method being analyzed. There are a number of factors to consider when deciding which parameters to track. Two of these are:

- The quantity selected to track performance must accurately reflect the waste you're interested in
- The quantity must be measurable with existing resources

When you conducted the baseline survey you determined the amount of waste generated on a macro scale. Then you used the flow chart and mass balance diagram to determine the waste-generation rates in a more detailed or micro scale for the selected waste stream. Measurements before the implementation are typically referred to as "baseline" measurements, while those after implementation are referred to as "actual" measurements. Projected measurements are those which you projected when you planned and designed the implementation. After deciding what data should be tracked, you will need to decide how to collect them and how to normalize the data.

When measuring the waste, it's a good idea to measure the weight of both the hazardous or toxic materials you put into your process as well as the waste stream emitted from the process. In fact, you may want to complete another mass balance diagram on this new system. The simple measurement of the weight of hazardous material fed into the process or waste emitted from the process may ignore factors that affect the quantity of waste generated; this is particularly true for your plants' production rate.

In general, waste generation is directly proportional to the rate of production of your finished product. When production of goods increases you expect your waste generation rate to increase as well, although perhaps not in a linear fashion. If you measure solely on weight, the effect of production increase or decrease could mask the effect your P2 implementation had on the actual generation of waste. Therefore, the attached worksheet at Figure 8.2 can be used to help you calculate the performance of the new system. This worksheet should be completed for the waste stream when it's baselined, for the proposed project, and for the actual implementation.

The method divides the weight of the material used (or waste generated) by the amount of finished goods manufactured by your process (e.g., gallons of product, number of cans, etc.). The calculation is labeled weight/unit product. Although this method of measuring waste reduction does take into consideration production rates, there may be other variables that cause you problems. These may include supply, quality, or plant maintenance.

Depending on the type and amount of waste you generate, you may have a considerable amount of information and data collected for your regulated waste streams. This information would have been gathered during the baseline survey and opportunity assessment phases. With this information and the determined tracking methods, you should be able to begin your measurement. Before you make your final decision I recommend that you also consider both the waste that is shifted to another medium and the toxicity of the waste.

- *Waste Shifted.* The pollution prevention option may eliminate part of the target material, but shift some of it to another waste stream or into another media such as from water to air. It is very difficult to track the shift of pollutant from one medium to another or determine if a new pollutant is created by the new process or the process change. Transferring the pollutant should be avoided if possible, but it is important to consider the relative impact on the environment.
- *Measuring Toxicity.* While I have discussed quantity produced as a measure of success, a reduction in toxicity is another measure of success. Of course the ultimate measure is reduced impact on the environment. I recommend you spot check toxicity levels as a regular measurement method.

Option Description_____

Baseline or Projected or Actual

1. Period Duration_____ From_____ To_____

2. Production per Period _____ Units _____

3. Input Material Consumption per period:

Material	Weight	Weight/Unit Product
_____	_____	_____
_____	_____	_____
_____	_____	_____

4. Waste Generation per Period:

Material	Weight	Weight/Unit Product
_____	_____	_____
_____	_____	_____
_____	_____	_____

5. Substances of concern- Generation Rate per Period:

Waste Stream	Substance	Weight	Weight/Unit Product
_____	_____	_____	_____
_____	_____	_____	_____
_____	_____	_____	_____

FIGURE 8.2   Waste stream performance evaluation worksheet.

TRACKING YOUR OA IMPLEMENTATION STATUS

Another way to track the success and status of each of your opportunities is to use a spreadsheet, such as the one shown in Figure 8.3.[2] This spreadsheet helps you keep track of all of your opportunities and the status of each of them. It should also become part of the documentation for your EMS. In order to complete this table, you need to understand the headings to each column. The description below provides the necessary explanations.

- *Box 1: Goal or Objective and Brief Description.* This information is found within your EMS. It should clearly identify exactly which goal or objective

1. Reduction Opportunity Goal and Description from EMS Plan	2. Scheduled Completion Date (Month/Year)	3. Completion Status: On schedule Y = Yes N = No C=Completed	4. Name of Toxic Substance(s); or Wastestream(s) Name; Include CAS #; and/or RCRA Waste Code #	5. State Volatile Organic Chemical "VOC", Ozone Depleting Chemical "ODC", "Both" or "N/A" For item in Box #4
Goal # ( ):				

6. If You Answered No No To Box #3, Provide Explanation(s) Including New Date(s):

FIGURE 8.3 P2 opportunity progress worksheet (complete this form for each opportunity approved by the environmental steering group).

7. State Briefly the Technology Used which Resulted in Achieving the Reduction (e.g. Replaced MEK based inks with water based inks)	8. Measured Reduction Quantity (Pounds or Gallons - circle one)	9. Month & Year Box #8 Was Measured	10. Baseline Year	11. Baseline Quantity (Pounds or Gallons - circle one)	12. Reduction Area	13. Reduction Method	14. Reduction Activity

FIGURE 8.3 (continued)

this action will meet and provide a brief description of the proposed implementation. You may also want to reference associated actions. Include what will be impacted as a result of reductions (i.e., reduction of air emissions, off-site waste management, etc.)

- *Box 2: Scheduled Completion Date.* Provide the original or proposed completion date, for each goal or objective as found in your EMS or in your proposal to the environmental steering group.
- *Box 3: Completion Status.* If the implementation is on schedule place a "Y" in the box. If delayed, place an "N" in the box. If the goal is completed place a "C" in the box.
- *Box 4: Name of Toxic Substance(s) or Waste Stream(s).* If reducing the use of a toxic substance, write the name of the toxic substance that is targeted for reduction, write the Chemical Abstract Number (CAS#) which may be found in the Material Safety Data Sheet (MSDS) or Form R. If reducing a waste stream, describe the waste stream and write in the name of the major toxic substance targeted for reduction within the waste stream. Write in the applicable RCRA waste code (i.e., F001, D001 etc.). The RCRA waste code may be found in your facility annual hazardous waste report document or on your manifests.
- *Box 5: Is the Toxic Substance or Waste Stream a Volatile Organic Compound (VOC) or an Ozone-Depleting Chemical (ODC)? (OPTIONAL)* If the substance in box number 4 is a known VOC or ODC, write "VOC" or "ODC" or "Both" within Box 5. Volatile organic compounds mean any compound of carbon, which participates in atmospheric photochemical reactions, excluding carbon monoxide, carbon dioxide, carbonic acid, metallic carbides or carbonates, and ammonium carbonate. Ozone-depleting chemicals include chlorofluorocarbons (CFC's), related bromine-containing chemicals called halons, methyl chloroform, and carbon tetrachloride.
- *Item 6: If You Answered No for Box #3, Provide Explanation(s) Including New Date(s).* If the stated reduction implementation is delayed, not on schedule, or was dropped as a goal, provide reason(s) for the delay or deletion and include the new projected completion date(s).
- *Box 7: State Briefly the Technology Used Which Resulted in Achieving the Reduction.* Provide *specific* information on how or what had to be done which did or will result in reductions as indicated in the goal. For example: install a gelcoat spray booth (to reduce emissions of styrene). Ensure that lids are provided and securely fastened on the acetone containers (to reduce acetone emissions). Purchase an acetone recycling unit (reduce purchases of acetone). Replace ethylene glycol with diethylene glycol. Replace 1,1,1-trichloroethane vapor degreasing with the Brulin 815 GD aqueous cleaning method (eliminate use of 1,1,1 TCA).
- *Box 8: Reduction Quantity.* Provide the quantity by which the toxic substance or waste stream was reduced since the reduction began. Indicate either pounds or gallons for the unit of measurement by circling the appropriate unit. Some reduction goals are measured only after completion, while others are not measurable in terms of quantities. Try to be consistent.

- *Box 9: Month Measured.* Provide in this box the month in which the reduction quantity (Box 8) was measured. For example, a facility began its pollution prevention activity or set its reduction goal in January, 1996 that resulted in a 50% reduction of TCA usage based upon a comparison of 1995. The 1995 purchase records were the measurement tool and were analyzed in January, 1996 and January 1997. The month and year measured was 1/96, then January 1997.
- *Box 10: Baseline Year for Reduction Measurement.* Provide in this box the baseline year used to measure the amount reduced (Box 8) (typically the year prior to the year the reduction goal began or was completed). For example, the pollution prevention activity began in 1996, and a 5000 pound or (50%) reduction goal was set based upon the previous years' hazardous waste report quantity. In this instance, the baseline year is 1995.
- *Box 11: Baseline Quantity.* Provide in this box the quantity of toxic substance used or waste stream generated during the baseline year indicated in Box #10. For example, if the pollution prevention activity began in 1997 and the baseline year is 1996, then the baseline quantity should be the quantity measured in 1996.
- *Box 12: Name the Reduction Area.* Provide the reduction area. The reduction area is one of the following categories: process (P), operation (O), waste stream (W), toxic usage (T), or other (O). If "other" is chosen, provide a brief explanation and include as an attachment. If the reduction goal falls within more than one category, please provide the one that most closely applies.
- *Box 13: Name the Reduction Method.* Provide the reduction method. The reduction method is one of the following categories: source reduction (S), recycling (R), treatment (T), energy recovery (E), or other (O). If "other" is chosen, provide a brief explanation and include as an attachment. Choose the category that most closely describes the method.
- *Box 14: Name the Reduction Activity.* Provide the reduction activity. The reduction activity is a subset of the reduction method. Determine which activity (i.e., good operating practices, spill control, etc.) most closely applies to the pollution prevention activity.

Another way to track successes is with trend charts, run charts, and bar charts that will graphically show the results of the reduction efforts. Other very effective tools to track the success of your P2 efforts include most of the tools I displayed in Chapter 5 on vulnerability assessment. You can apply the bar charts, pareto charts, and run charts to describe the amount of waste generated over time. Hopefully, the trend will indicate a reduction in waste generation. When you are gathering data to make your measurements, be sure to use the same measurement system you used when you conducted the baseline assessment I discussed in Chapter 6. Recall from your earlier readings that I suggested you develop a standardized spreadsheet, checksheet, or electronic "clipboard" to assist you in this labor-intensive stage. Taking the time to develop a standardized, simplistic form or other tool will ensure that you get consistent and reliable data with minimal labor input. Invest the effort to standardize this system!

CIP STEP 8. STANDARDIZING THE SOLUTION

After implementing a couple of P2 successes it should be easier to keep the program going, but quite often leadership begins to lose focus. By integrating the P2 program into the EMS, it, of course, keeps the focus high. Another way to keep interest high is by tracking the success of P2 efforts along with the EMS goals and objectives. The metrics and measures are the key to success in this business. Remember, the models I've shown are continuous improvement models. You keep going through these cycles again and again until you achieve your objectives. Keep your focus on P2 initiatives until you have exhausted all simple and low cost initiatives. Then move into the more formal EMS steps discussed below.

ASSIGN ACCOUNTABILITY FOR WASTES

Operating units that generate wastes should be charged with the full costs of controlling and disposing of the waste they generate. The full costs should take into account the indirect costs, such as potential liabilities, compliance reporting, and oversight. Burying the waste management costs in general overhead can reduce the impact of these costs and therefore not motivate the process owners to attempt to reduce the cost. Every single waste stream should have a single individual responsible for that waste stream. They should track the amount generated, be responsible for all compliance activities, and be responsible to assist the P2 team find better methods to reduce the waste generated. These individuals are probably already identified and working as part of your existing compliance programs but you need to ensure that they:

- Have the appropriate training
- Have the necessary equipment and supplies
- Know emergency response actions if relevant
- Have the appropriate level of compensation and authority to conduct their assigned tasks

TRACKING AND REPORTING

During the baseline survey stage I discussed the need to determine the exact amount of waste that you generate from all waste streams. In the P2 strategic plan section, I provided some objectives to help you measure your success. To continue to track your waste generation rates, you will need an information management system to track and retain the data necessary to measure P2 program results. You will need to ensure that these data are reviewed and reports are prepared at meaningful intervals. The reports should be generated frequently enough to enable managers to monitor and adjust their programs to adhere to the objectives established during the EMS planning stage and provide feedback to their staff.

Quarterly Program Review

Senior management should hold quarterly environmental leadership meetings as discussed in Chapter 3. At these meetings, reduction trends toward goals and objectives should be presented; then the reviews should be communicated to all employees. Also, program success should be recognized and any changes in objectives or policies announced and explained. If you have slippage in your program schedule, senior management needs to be aware of this so they can allocate additional resources or modify the schedule.

Training

One of the basic elements of a mature EMS and a P2 program is employee training. It should involve all levels of personnel within the organization. In fact, small tailored training is the most effective. The actual objective of the training should be to make employees aware of their impact on waste generation and how to recommend ways to reduce the waste. It should include relevant information for employees to understand how they can help achieve the EMS policies, goals, and objectives. The training can be phased such as:

- Generic new employee orientation training
- Advanced/process specific training
- Annual retraining
- Raise awareness of which processes within your facility impact the environment
- Inform your employees of specific environmental issues
- Train employees in their P2-related responsibilities
- Recognize employees for their hard work
- Publicize success stories

Increase Internal Communication

A mature EMS has complete communication and information-sharing systems. These could include such activities as:

- Regular status reports on improvements
- Clearly defined objectives
- Solicit and follow up on employees' suggestions
- Employee rewards program
- Environmental impacts as part of annual performance reviews

CLOSING THOUGHTS

Once you have become fully comfortable with the implementation of your opportunity and you have standardized it across all similar processes, then it's time to "turn the wheel" and work though another opportunity. Go back to step 5 of the P2

CIP and identify another opportunity and follow through implementation and standardization of reasonably simple opportunities that provide favorable return on investment. Once you feel you have reduced all easily reduced waste streams, then it is probably time to review your EMS policy and goals again and select more challenging reduction opportunities. Once your waste-generating processes are reasonably under control and reduction efforts are successful, then you may want to consider auditing your EMS to determine the health of your program and determine if certification is a possibility.

REFERENCES AND NOTES

1. Adapted from a public document by Pingenot, J. and Voorhis, J., *The Pollution Prevention Assement Manual for Texas Businesses,* prepared for the Texas Water Commission, 1995.
2. Worksheet and description adapted from U.S. Air Force internal publication entitled *Pollution Prevention Management Action Plan,* 1995. I understand the original worksheet was developed by the State of Arizona, Department of Environmental Quality.

9 Your EMS Manual and P2 Plan

Within the ISO 14000 standards, there is some discussion on the development of an EMS manual which rolls up much of the information we've discussed so far in this text. In this chapter, I wanted to provide you with an example which you could use as an outline to help you develop your EMS manual and your P2 implementation plan. Although the final EMS manual will probably take a different shape, this example is a good first step in developing yours. Then it can evolve to become more comprehensive as your EMS matures. Also, as you develop your first plan, remember that it's actually the process of planning and thinking through your strategies and objectives which provides the most value to your organization. The actual plan is simply the end result of that effort. Of course, you must stick to the plan you've laid out to achieve the results you expect from all of this effort.

The strategic planning process you used to develop your goals and objectives for other organizational initiatives should also be used to develop your EMS manual. In fact, the EMS should link to other strategic planning initiatives through the policy statement, key focus areas, and goals. As discussed earlier, P2 goals and objectives are integral to your EMS. In fact these are the predominant part of your EMS in a proactive organization. Then, once you have tackled your easiest P2 challenges and your environmental processes are stable, you can move forward to other parts of an effective EMS, such as focusing on completing an audit, training more employees, or enhancing external communication.

The EMS manual should contain the overall organizational policy and clearly define how it will be implemented as we discussed in Chapters 2 and 3. The implementation discussion should include an organizational layout showing who is in charge of various phases or parts of the implementation. It should also detail how various requirements of the standard will be tackled throughout the organization.[1] The manual will eventually become a very comprehensive document. But don't duplicate effort. If you currently have documentation which I suggest should be part of your EMS manual, don't rewrite it as part of your EMS — simply reference it in your EMS. The value of developing an EMS manual includes:

- Having a single document which includes all environmental policies, goals and strategies;
- Having clarified and written responsibilities for virtually all levels within an organization;
- Consolidating and indexing all environmental information in a single location which is useful for continuity among employees, conducting audits of your processes, and a great tool for training new employees;

Your EMS manual should be constructed around the requirements of the standard or code you're implementing. In Chapter 2, I discussed the requirements of the ISO 14000 EMS, and, as such, a manual should clearly describe how your organization meets each of the areas required under the ISO 14001 standard. I recommend that you turn back to Chapter 2 and review the paragraphs on ISO 14001, especially Implementation and Operation and Figure 2.9 and the paragraphs on Measuring and Corrective Action plus Figure 2.10 as you consider developing or constructing your EMS manual. These paragraphs discuss the requirement under ISO 14001 and as such should be covered in your manual. Once these are understood, then go further by fully developing your goals, objectives, and strategies for implementation. I believe the basic sections of your EMS manual should include the following.

- Your organizational environmental policy.
- The structure of your organization to implement the policy. I'd include the roles and responsibilities of individuals and teams, such as the environmental steering group, P2/ISO 14000 teams, and supervisors or managers, in implementing the policy as we've discussed already in this text.
- Identification of the mandatory regulatory requirements at the federal, state, and local levels. This could be done simply by citing the regulation and describing where to find them within your organization.
- Description of your strategy for meeting your policy via implementing and maintaining your EMS. Show your long-term and shorter-term strategies, using such tools as Gannt charts and timelines. Include your overarching goals and the appointed OPRs for each goal, indicating when the goal will begin and end. This will be a very dynamic section. Don't be afraid to change it or miss a milestone. Its purpose is to provide a road map and indicate a vision of where you're going and how you'll get there. It should never constrain your actions.
- Description of your goals, objectives, strategies, and measures. This grows more specific from the previous section on your strategy. As you begin your EMS implementation, this is the key portion of your EMS manual because the clearly defined objective will drive action. Later in this chapter you will find numerous examples to help you develop this critical part of you EMS manual.
- List of any specific requirements outlined by the code or standard you're implementing that are not already covered.
- Description of your environmental training plan. Your strategy for worker training including the who, what, when, where, and why of each type of training. Include a brief description of each type of training you provide and explain where to find the appropriate lesson plans and or training material.
- Discussion of your EMS documentation control. Your EMS manual needs to have a detailed explanation of how you file and catalog your environmental records. This includes information from disposal manifests to training logs to plans and permits. ISO 9000 audits constantly identify document control as an area needing significant improvement. One area

that needs constant attention is to ensure that you are using the most current or recently revised document. This is especially true with operational procedures. There should also be a process in place to ensure that documents are either reviewed and updated or disposed of at appropriate intervals. You have to establish and maintain procedures for identification, collection, indexing, filing, storage, maintenance, and disposition of environmental documents and records which should be readily available and protected against loss. Your document control procedures should include:

- The document retention time. How are documents filled, indexed, and how long are they kept before disposal?
- Distribution of documents. How are documents sent to users? Is it a systematic process with some type of controls or is it unstructured and haphazard? How are revisions sent?
- Procedure for developing documents. Are there organizational standards? What are the approval procedures?
- Discussion of the management of your environmental records. Records are the detailed documentation that identifies your procedures to show that you are in compliance with your EMS. These are typically standardized forms or letters which are used as proof of actual improvements or standardized procedures during an audit. Although you would not keep these as part of the actual EMS manual, they should be referenced in the manual along with the location where they can be found. Some records you should consider keeping in addition to those required by compliance related laws include:
  - Training attendee lists
  - System failures and the corrective actions you've taken or plan to take
  - Changes in procedures as a result of process analysis
  - Results of audits
  - Reports from vulnerability assessments, baseline surveys, and opportunity assessments
  - Process measurements and any reduction trend and run charts or other data.
- Discussion of how you are achieving operational control of those processes which affect the environment. Describe the activities, processes, and operations which impact the environment and show how your goals and objectives target these particular processes. If you were effective in your strategic planning and vulnerability assessment sessions, you have already completed this action. Simply document your previous thoughts and actions in your EMS manual. Also show how your P2 implementation activities and any compliance activities have helped to minimize your impact to the environment. Eventually this section will grow from a list of processes which affect the environment to become a detailed analysis showing how your activities have reduced your impact on the environment. I envision that this chapter will probably show process flow charts, reduction trend charts, or other measurement data presentation tools showing

how your efforts are achieving success in stabilizing these processes and reducing your impact on the manmade and natural environment.

- Discussion of how you conduct emergency response. The specifics of how you prepare for and respond to environmental emergencies must be clearly outlined by your organization. This would include your specific checklists, equipment inventories, and any tests or practice sessions you may have. This information is probably already required by environmental laws such as RCRA, or EPCRA and the Clean Water Act. So if you've already developed these procedures and processes and otherwise meet these requirements, simply reference them in your EMS manual. Eventually, as you reduce the toxicity of the materials used in your processes, you may be able to eliminate the need for having emergency response requirements.
- Emissions and Waste Treatment. This area is fully covered by regulatory requirements and various permits under separate environmental media requirements. You simply need to reference and recognize those actions in your EMS and reference the appropriate documentation.
- Audit System. You simply need to describe the strategies for auditing your EMS. I'll discuss audits in more detail in Chapter 10.
- Management Review. Describe your strategy for senior management review. Be specific here. Include the who, what, when, where, and why's.

The EMS manual should be a very dynamic and constantly changing "working" document. The purpose is to provide a management tool to assist you to stay on track with the process of reducing the hassles of the environmental business. It is a product which should tie together the loose ends of various compliance based programs. But the heart of it is the objectives which should keep you focused on improvements to your systems in an effort to meet your environmental policy.

EMS STRATEGIC PLANNING

Before I dive into an example, I want to reintroduce you to the strategic planning model and terms I used in Chapter 3. The planning model I like to use is best described by a pyramid, as shown in Figure 9.1. The relationships of the various parts of the pyramid describe their logical relationships beginning with a broad visionary policy statement and increase in specificity through key focus areas (KFAs), goals, objectives, strategies, and measures.

Another model is the "waterfall chart", as shown in Figure 9.2. This model also shows the relationship among the various elements. Both models help you keep focused on the various parts of the planning process; this is important to achieve your objectives. The identified goals are the focus to which all improvement areas should be aligned. Then the goals are further focused through objectives and strategies.

THE POLICY STATEMENT

As discussed in Chapter 3, the policy statement is the heart of your EMS and P2 activities. The policy statement is developed by your senior leadership and should articulate their vision as to where they want the environmental program to go.

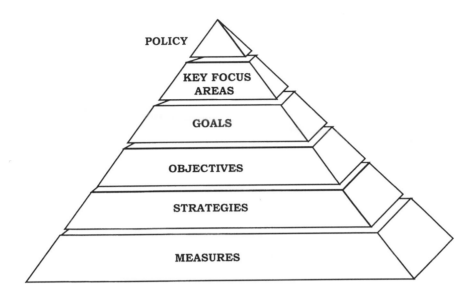

FIGURE 9.1   The EMS policy deployment pyramid.

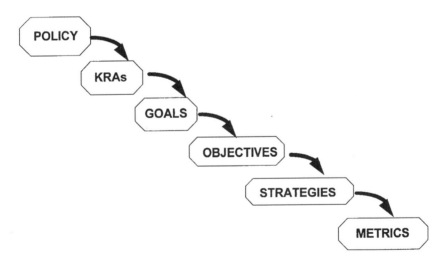

FIGURE 9.2   The EMS policy deployment "waterfall" chart.

KEY FOCUS AREAS (KFA)

These essentially describe those few areas where you want to focus your energies to get the most results from limited resources. What are those major categories of customer requirements critical to the success of your organization? What are those major environmental requirements which are causing you problems or draining your resources the most? Chapter 5, the Vulnerability Analysis Assessment, provides screening tools to help determine those key areas where you may want to expend your energies. Identifying your key focus areas is an optional step in the strategic planning process if you have only a couple of areas you plan on setting goals for — but if you have many areas, you may need to develop your KFAs in order to focus your operation. KFAs are usually very general in nature. Some examples may include: reducing waste, increasing training, increasing communication, cost reduction, personnel development, process improvement, or waste reduction by types of waste or media (hazardous waste, solid waste, air emissions), environmentally safe purchasing, public awareness, or maintaining compliance.

GOALS

Goals are the specific areas where senior leadership wants you to place your efforts subordinate to the KFAs. Separate goals should be set for both your reduction as well as your improvement efforts. ISO 14000 refers to this section as "objectives," while typical TQM or management material refers to these as goals. Whatever you decide to call them, it's important to note that they are relatively specific descriptions of activities you or your senior leadership would like to achieve. Typically, goals are stated as phrases such as:

- Promote sound waste management practices.
- Minimize the use of hazardous materials.
- Reduce air emissions to the lowest practical level.
- Improve operational performance through employee training.

OBJECTIVES

Objectives are more specific than goals and typically force some type of action. Objectives must be measurable over time and should also have a "sun down" or completion date to them; this is the date you expect to achieve your objective. ISO 14000 documents refer to these as "targets," which is a good term because the results are what you are shooting for. Objectives should be clearly written, concise, and results focused. Examples include:

- Reduce solid waste by 30% within 3 years.
- All paper products purchased shall contain at least 20% postconsumer recycled content.

STRATEGIES

Strategies are the actual steps you will go through to complete the objectives. In the examples provided, I have asked questions for you to consider when developing your strategies. Hopefully, these questions will lead you to develop the strategy which is most appropriate for your organization. The strategies can become the plan or the beginning of the plan on how you'll achieve your objectives and goals.

Measuring Success

In order to clearly communicate the effectiveness of a P2 program, it is necessary to first identify the amount of waste generated prior to P2 implementation (a baseline), then measure after the initiative is implemented to ensure that you are meeting your stated objectives. An "owner" must be identified for every waste stream, process change, and initiative in order to measure the improved process and achieve the objectives. An example of how this all fits together is shown below.

SOME EXAMPLES

The next pages are examples of KFAs, goals, objectives, strategies, and measures which I have seen used to help focus efforts on P2 activities within an organization. I am not advocating that you try to focus on all of these goals at once; there are too many presented here. I suggest that you pick three or four goals for your organization and focus on those. Do not attempt to do more then that.

KFA 1. Hazardous Waste Reduction

Goal 1.1. Quantify Hazardous Waste Generation

Objective 1.1.1. By the end of this year, determine the amount of hazardous material being purchased and the amount of waste generated from 100% of your processes.

Strategy. Conduct a vulnerability assessment and baseline assessment or survey to quantify the amount of hazardous waste generated and determine exactly where it comes from. Subprocesses or tasks may include:
- Conducting the baseline survey.
- Developing a database for waste tracking and inputting the data into the database.
- Verifying that a baseline exists for hazardous waste (HW).
- Verifying that all existing processes and systems which use hazardous materials or generate HW are identified.

Metrics.
- Hazardous waste disposal in pounds.

Data Presentation. Since a baseline survey is a static value, by definition it does not change over time, but it is necessary to have this information for trend analysis and to identify opportunities for reduction. The baseline data may be tabulated in the format shown in Figure 9.3 and tracked over time in a trend chart or a run chart.

Goal 1.2. Identify opportunities for waste reduction.

Objective 1.2.1. Conduct an opportunity assessment (OA) on 100% of the major waste streams this year.

Strategies. Implement the OA process then evaluate effectiveness.
- Obtain a consultant or identify an internal team to conduct the OAs.

WASTE STREAM	POUNDS GENERATED or PURCHASED
Haz Waste	
Solid Waste	
Air Emissions	

FIGURE 9.3   Baseline data table.

- Have the OA team review information from baseline surveys and discuss information with process owners.
- Visit processes and conduct "round-table" discussions to get good ideas from waste-generating activities.
- Review draft reports and provide comments. Consultant completes report.
- Identify a process owner for each OA.
- The process owner should review and approve OA recommendations prior to cross-feed status to the P2 working groups.
- Verify that an OA is conducted at least every 3 years on all processes which contribute to at least 15% of a waste stream.
- Verify that each OA provides specific recommendations on waste stream reduction.
- Verify that OA recommendations are validated or approved by the "user agencies" and/or the P2 working group.

Metrics.
- Total number of waste streams divided by the number of OAs conducted expressed as a per centime.
- Number of valid or implementable OA recommendations.

Objective 1.2.2. Develop action plans to implement 10% of all OA recommendations by the end of the year.

Strategy. Develop a "road map" for the implementation, then implement the opportunity assessment recommendations.
- Gather inputs from the P2 working group and the OA recommendation's process owners.
- Develop an implementation plan.
- Identify a process owner for each action.
- Does each OA specifically identify equipment and projects or process changes needed to meet reduction objectives?

Metric. Number of OAs implemented
- Number of waste streams/number of OAs conducted shown as a percentage.

- Implementation of valid OAs within 6 months of completing the action plan.
- Evaluate effectiveness of OA within 1 year of implementation.

Goal 1.3. Implement a Hazardous Material Tracking System

Objective 1.3.1. Track 100% of hazardous material purchases by the end of the calendar year.

Strategies. How does the function which uses hazardous material maintain inventories of material purchased and used as required by SARA Title III and public health driven requirements?

- Inventory all existing processes and systems which use hazardous material in an effort to manage hazardous material so that it contributes to reducing hazardous material purchase and hazardous and solid waste disposal.
- Centralize hazardous material procurement, distribution, and reuse.

Metrics.

- The amount and type of hazardous material purchased each year.
- Percent reduction in hazardous material purchases.

Goal 1.4. Reduce Hazardous Material Disposed of as Hazardous Waste

Objective 1.4.1. Reduce expired shelf life items to 2% of the organization's total hazardous waste disposed of.

Objective 1.4.2. Reduce the disposal of unused hazardous material to 10% of the total amount of hazardous waste disposed of.

Objective 1.4.3. Reissue at least 90% of unused hazardous material.

Strategies. Reduce hazardous waste disposal volume and toxicity through reduction in the unnecessary use and purchase of hazardous material.

- Ensure that you track the disposal of unused or partially used hazardous materials.
- Verify that expired-shelf-life items disposed of as hazardous waste are under 2% of the total hazardous waste stream.
- Verify that purchases of the EPA 17 toxic chemicals are reduced by 50% from the baseline by the end of the calendar year.
- Verify that the organization has a process in place to collect and reissue unused hazardous material.

Metrics.

- Percentage of hazardous waste stream reduced as compared to the baseline.
- Percentage of hazardous material reused or reissued.

Data Presentation. Tracking the reduction trends over time is a good way to show success in this program. This can be accomplished graphically or on a spreadsheet as shown below. Figure 9.4 is a hazardous waste reduction table, while Figure 9.6 is a proposed spreadsheet for tracking reduction trends.

INITIATIVE	COST SAVINGS (To Date)	% OF HW STREAM	% REDUCTION OF HW STREAM
Expired Shelf Life			
Reuse/Reissue			
Solvent Recyclers			
Parts Washers			
Antifreeze Recyclers			
Other Equipment			
Solvent/Cleaner Substitutes (list)			

FIGURE 9.4   Hazardous waste reduction table.

A trend analysis of each major waste stream should be developed and presented to senior leaders at the environmental steering group or at a sub-working group. Every hazardous material used should have an "owner" attached who is responsible for the reduction measures and trend analysis.

Goal 1.5. Reduce Hazardous Waste Disposal

Objectives 1.5.1 By the end of next year, reduce the hazardous waste disposal amount by 25% per year as compared to the baseline.

Strategies. Accurately document hazardous waste generation rates from specific processes and track reduction trends using a database. Some questions you can consider include:

- Verify that you have hazardous waste data sorted by various processes.
- Verify that someone compares current hazardous waste generation rates to the baseline data to track reduction trends and identify opportunities for reduction.
- Verify that the environmental steering group is aware of the current hazardous waste reduction status. Examples of how this could be accomplished include briefing charts, and background papers.
- Verify that you're recycling hazardous waste whenever possible. This can be accomplished by using solvent stills and antifreeze recycles.
- Analyze service life of hazardous material and identify ways to extend the material's life such as with additives or storage conditions.
- Establish award incentives to recognize programs or individuals that reduce use of hazardous material.

HW Process Code	1996 Baseline	1997	1998	1999	Reduction Trend
HW Total					
AB					
AC					
BA					
BO					

FIGURE 9.5 Hazardous waste disposal trend analysis table.

Metrics.
- Amount of HW disposed of by type of waste.
- Reduction trend compared to baseline.
- Amount of HW recycled.

Data Presentation. Tracking the reduction trends over time is a good way to show success in this program. This can be accomplished graphically or on a spreadsheet. Figure 9.5 is a proposed spreadsheet that could be used to help track reduction trends. A trend analysis of each major waste stream should be developed and presented to senior management at the environmental steering group or at a working group. Each waste stream should have an "owner" attached who is responsible for the reduction.

KFA 2. REDUCE AIR POLLUTION

Goal 2.1. Reduce Air Emissions

Objective 2.1.1. By the end of 1999, reduce the VOC emissions by 50% from the baseline.

Strategies. Document air emission's generation rates from specific processes, reduce emissions and track reduction trends.
- Provide a database to track and calculate air emissions.
- Compare current air emission generation rates to the baseline data to track reduction trends and to identify opportunities for reduction.
- Ensure that the environmental steering group is made aware of current emission reduction status. Examples of how this could be accomplished include briefing charts and background papers.
- Establish a baseline for volatile and other air emissions.
- Characterize all waste streams released to all media.
- Implement a program to reduce volatile air emissions by 50% from the baseline within 5 years.

- Implement a program to reduce total releases of industrial toxics by 50% within 5 years.
- Identify and implement pollution prevention measures to minimize releases of volatile organic compounds and industrial toxics to air and water.
- Ensure that pollution prevention opportunities are exhausted before end of pipe solutions are used to control releases to air and water.

Metrics.
- Amount of air emissions reduced by each process.
- Reduction trend compared to baseline.

KFA 3. ELIMINATE UNNECESSARY STORM WATER CONTAMINATION

Goal 3.1. Reduce Nonpoint-Source Pollution

Objective 3.1.1. Reduce nonpoint-source pollution to near zero within 5 years.

Objective 3.1.2. Eliminate the use of ethylene-based runway and roadway deicers within 2 years.

Objective 3.1.3. Inventory all oil–water separators this year.
Strategies. Effectively reduce the release of pollutants to waterways, storm drain etc. Some steps you may want to consider include:
- Provide a contractor to inventory location of all oil and water separators.
- Provide guidance to process owner of each oil and water separator.

Metrics.
- Amount of discharges reduced by each process.
- Reduction trend compared to baseline.

KFA 4. SOLID WASTE REDUCTION

Goal 4.1. Reduce Dependency on Landfills

Objective 4.1.1. Determine the amount of solid waste being generated by various waste streams by completing baseline assessment this year.

Objective 4.1.2. Reduce MSW disposal by 50% from baseline within 5 years (10% reduction per year).

Objective 4.1.3. Recycle 25% of the total waste stream by next year.
Strategies. Reduce the amount of municipal solid waste (MSW) going to landfill through source reduction, recycling, and composting.
- Implement active recycling programs to include: office recycling program, cardboard containers.
- Ensure that all offices, the printing shop, clinics, and other areas that generate a great deal of computer and bond paper recycle their paper.

PRODUCT	AMOUNT (This Qtr)	PROFIT /COST	AMOUNT (Ave Qtr)	% of MSW STREAM	% REDUCTION IN MSW
Curbside					
Office Paper					
Computer Paper					
Cardboard					
Glass					
Plastics					
Alum. Cans					
Metals					

FIGURE 9.6 MSW reduction trend spreadsheet.

- Implement active composting programs if your organization generates a lot of food wastes or yard wastes.
- Compare current MSW generation rates to the baseline data to track reduction trends and to identify opportunities for reduction.
- Is senior management aware of current MSW reduction status? Examples of how this could be accomplished include briefing charts and background papers.
- Does your MSW disposal service contract payment schedule based on the amount of waste disposed? *Note:* Many contracts are paid by the number of dumpsters picked up, not by weight or volume. Since volume and weight of trash will be reduced as recycling increases, don't forget to adjust your dumpster pick-up to save funding.
- Sell, promote, and support awareness of solid waste reduction and recycling by establishing an incentive/award program for best recycling program/individual.

Metrics.
- Amount of material recycled, composted.
- Percent reduction in MSW stream.
- Reduction in solid waste disposal costs.
- Cost of MSW reduction efforts.

Data Presentation. Tracking the reduction trends over time is a good way to show success in this program. This can be accomplished graphically or on a spreadsheet as shown in Figure 9.6.

Goal 4.2. Purchase Environmentally Safe Products

Objective: 4.2.1. At least 20% of all nonpaper products and 50% of all paper products will contain recycled content material.

  Strategies. Purchase recycled content materials to create a market for the products we recycle and encourage personnel to use products containing recycled materials.

  - Develop clear guidance for procurement officials.
  - Direct the use of recycled photocopy paper, toner cartridges.
  - Direct the use of retread tires.
  - Verify that all photocopy paper purchased is recycled content paper.
  - Verify that at least 50% of all paper products purchased contain recycled material.
  - Verify that at least 10% of all nonpaper products contain recycled material.
  - Verify that the internal newsletter is printed on recycled newsprint.
  - Verify that the organizational stationery is printed on recycled letterhead paper and that it has the recycled logo.
  - Verify that senior management is aware of current affirmative procurement trends. Examples of how this could be accomplished include briefing charts and background papers.
  - Facilitate the procurement of recycled/recycle content items by searching out opportunities for substitution of existing items with recycle materials where appropriate.
  - Establish affirmative procurement programs so that 100% of purchases of products meet EPAs designated guideline standards.
  - Increase success of recycling programs by stimulating market demands for recycled products by purchasing these products whenever possible.

  Metrics.
  - Amount of recycled content paper purchased.
  - Amount of recycled content "nonpaper" products purchased.

  Data Presentation. Tracking the reduction trends over time is a good way to show success in this program. This can be accomplished graphically or in a table. Note that no trend line is required because the "baseline" is constantly changing. The objective is to have at least half of all paper purchased at any time to be recycled content paper; therefore the trend line is horizontal or steady state.

CLOSING THOUGHTS

The EMS manual is a very important document which links together the numerous policy statements, goals, management plans, permits, records, training reports, and other formal documentation required to effectively lead an environmental management program. ISO 9000 audits typically show that documentation is the weak link in achieving certification. I'm sure this trend will continue with ISO 14000 because the sheer volume of documentation required by environmental regulators is so great.

As you have learned, an effective EMS should be based on pollution prevention and continuous improvement. With this principle in mind, you should first focus on reducing pollution through concrete, measurable objectives such as those presented within this chapter. These objectives and the supporting goals and policy statement become the first part of your EMS manual. Focus on reducing your compliance burden by having a very effective prevention program; then, once your prevention program is mature, worry about completing the rest of your EMS manual and developing a full EMS in accordance with the code or standard you're following.

REFERENCES AND NOTES

1. Most of the discussion on the EMS manual came from a paper written by Marcos Oliveir, Documentation for Environmental Management Systems, presented at the International Conference on Quality in Yokohama Japan, Oct. 1996. See http://www.ISO14000.com for a copy of this paper.

10 Checking Your Environmental Pulse

Once you have stabilized your environmental processes, implemented your EMS, conducted a significant amount of training, and successfully completed numerous P2 initiatives, you may consider checking the health of your EMS. This includes measuring your environmental processes and conducting an EMS audit. The purpose of this type of audit is to check the health of your EMS and to ensure that your organization is conducting business as you stated it should when you developed your policy, goals, and objectives. This EMS audit, sometimes referred to as an eco-audit, is a more holistic and proactive evaluation of your environmental impacts and how senior management responds to those impacts than other types of audits.

As a refresher from earlier discussions in Chapter 3, the EMS audit and performance evaluations essentially fall into Step 5 (Measurement and Evaluation) of the EMS implementation process as shown in Figure 10.1. They are also represented in Step 7 (measurement) of the P2 continuous improvement process as discussed in Chapter 4 and shown again in Figure 10.2.

In the past, environmental audits focused on four separate and distinct areas; compliance audits, waste audits, product audits, and energy audits. Typically, environmental compliance audits are the most common. These audits evaluate whether or not your processes are within full compliance with federal, state, and local environmental laws, regulations, and codes. Waste audits are very similar to the baseline and opportunity assessments discussed in Chapters 6 and 7. These typically analyze existing waste streams to identify potential for reduction, recycling or reuse. Product audits examine the impact a product may have on the environment during its useful life and after disposal. Energy audits are used to determine if unnecessary wasting of resources is taking place. These usually occur as a joint partnership with the energy supplier and the customer, or energy user, who work together to identify opportunities for reduction.

The EMS audit goes further than the other audits. It is an evaluation of your systems to ensure conformance (not performance) to the EMS standard. In essence it's a verification or validation. Although you will probably continue to audit your systems to ensure that they are in compliance with environmental regulation, you can use your EMS audit to check to see that your systems are meeting your EMS policy, goals, and objectives, and to assess the gap if they are not meeting those stated objectives. You can also check to ensure that you're reducing pollution and therefore your impact on the environment. You use the EMS audit to assess the health of your environmental systems just as you would use the Malcom Baldridge criteria to measure the health of your quality management systems. Once you are

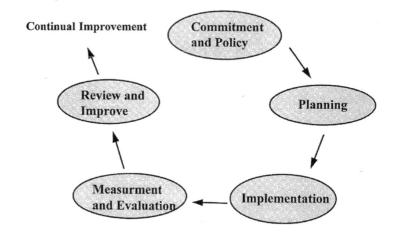

FIGURE 10.1   The EMS continuous improvement cycle.

FIGURE 10.2   The P2 continuous improvement model.

very happy with the health of your environmental management processes, then you may consider using an external agency to conduct an audit with the objective of certifying your EMS.

Some of the reasons you may want to conduct an EMS audit include:

- It helps diagnose your strengths as well as identify areas for improvement.
- It promotes a sharing of best practices and processes.
- It provides a framework for strategic planning.
- It provides a road map to continuous improvement.
- It provides quantifiable measurements.
- It provides a framework to identify and share strengths.

A BRIEF OVERVIEW OF COMPLIANCE AUDITING

National efforts to assess the health of an organization's environmental compliance program led to the development of self-auditing programs. The concept behind environmental compliance audits is simple — find and correct problems early to save time, money, and hassles later. The early detection of improper procedures allows time for correction before more expensive cleanup actions or regulatory agency involvement is required. Investment into compliance oriented programs, such as auditing, reduces the need for resources to be expended in penalty-type payments (fines and cleanup costs) in the future. Environmental compliance audits are one of the most effective tools for ensuring compliance. Typically, there are four goals of an environmental compliance auditing program:

- To evaluate operations for compliance to statutory requirements and taking action to correct where needed
- To improve environmental management by having a single program that addresses all efforts for achieving, monitoring, and maintaining compliance in a single program
- To create a standardization of methods and increase general awareness of environmental responsibilities
- To avoid financial or criminal liability

Typically, successful compliance audit programs have several common attributes. These include:

- Establishing minimal standards
- Mandating use of the standards organization-wide
- Maintaining currency of the established standards
- Providing training over a broad spectrum of workers and managers
- Establishing a uniform reporting system
- Having top level management involved
- Assuring management responds to prioritized needs to maintain compliance
- Establishing a cycle of environmental audit processes

The EPA continually stresses the importance of compliance auditing, with most organizations agreeing to and adopting the concept. The EPA uses two definitions for environmental compliance auditing. The first is "a systematic, documented, periodic and objective review by regulated entities of facility operations and practices related to meeting environmental requirements."[1] The four basic tenets — systematic, documented, periodic, and objective — are vital to making an audit successful. The second definition is "an independent assessment of the current status of a party's compliance with applicable environmental requirements or a party's environmental compliance policies, practices and controls."[2] Both definitions focused on using the audit to determine the level of compliance with environmental requirements.

The Government Accounting Office's (GAO) 1995 environmental auditing report found that organizations that used environmental auditing reaped benefits such as reduced liability and cost savings. GAO also found several common characteristics of effective auditing programs. These characteristics included:

- Top management provided support and the resources to hire and train personnel
- Audits were independent from external or internal pressures
- Audits employed quality assurance procedures to ensure accuracy and thoroughness[3]

Overall, organization wide compliance audits have been extremely successful in ensuring that organizations meet environmental laws and regulations. They also have:

- Improved environmental management worldwide
- Improved environmental compliance management in the U.S.
- Helped support financial programs and budgets for environmental compliance
- Ensured that environmental programs are effectively addressing environmental problems that could:
  - Impact organizational effectiveness
  - Jeopardize the health or safety of personnel or the general public
  - Significantly degrade the environment
  - Expose the organization and its people to avoidable financial liabilities as a result of noncompliance with environmental requirements
  - Erode public confidence in the organization
  - Expose individuals to civil and criminal liability
  - Anticipate and prevent future environmental problems[4]

While compliance audits have been very effective in achieving compliance, they are not a holistic approach to measuring the overall health of your organization's environmental stewardship. A compliance problem identified during an audit is probably only a symptom of a process problem. Since compliance audits look at subprocesses or tasks, such as whether permits were properly completed or whether hazardous waste containers are correctly labeled, they do not identify the root cause of the noncompliance issue. Typically the actions taken after the audit are to close the findings. Since the findings are at the task level, the system creating the finding is not corrected and the noncompliance continues.

Another weakness of compliance audits is that they are labor-intensive and do not promote continuous improvement. I have observed compliance audit teams of over 25 people conducting annual audits. Typically, compliance audits are conducted by external organizations that can cause the organization being audited to overreact to an upcoming audit. Many senior managers do not want a "black-eye," so they "clean up for the audit." This mentality yields more waste, distorts the true compliance picture, and sends a signal to employees that environmental issues are only important prior to an audit. Audits of your EMS are a more complete assessment of the health of your program including your management practices.

EMS audits and compliance audits can be integrated. In fact, when you conduct an EMS audit, you should also review your compliance auditing program to ensure that the system you've established to produce compliance is properly functioning. Let's now take a look at the requirements under ISO 14000 regarding performance evaluations and auditing. After that we'll discuss how to execute an EMS audit.

ENVIRONMENTAL PERFORMANCE EVALUATIONS

ISO 13031 is the environmental performance evaluation (EPE) standard that can be used to help you evaluate your performance against the goals and objectives you set during your planning process. Although EPE is not part of auditing your EMS, it is a management tool to evaluate its health. Remember where we are on the EMS model: we're to the point of measuring and evaluating our efforts so far. The EPEs can be used to help you with this step in the cycle.

Recall from Chapter 2, the ISO 14001 EMS standard requires an organization to have a method in place to oversee how well you're doing at meeting the goals and objectives you have set for yourself. Then, in Chapter 3, we discussed how to develop those goals and objectives and how to set targets. The EPE standard essentially discusses the same information, developing detailed process measures and outcome measures for all processes that impact the environment, then tracking their progress over time. In TQM terms, these may be called process metrics or quality performance measures (QPMs).

The example goals and objectives provided in Chapter 9 are excellent pollution prevention or waste reduction indicators. They included such things as:

- Amount of waste reduced over time
- Reduction of air emissions over time

Some other indicators could be more product-specific, such as amount of hazardous waste generated per pound of product generated or tons of air emissions per amount of energy used. Still other indicators could be related to your communications and training programs.

Internal performance evaluations should not be confused with conducting an audit. The EPE is an ongoing measurement of those processes that impact the environment. Usually data are gathered and analyzed by line employees who are part of the production process. This is quite analogous to statistical process control charts and other measurements and data gathering tools used for process improvement under ISO 900 and other TQM methods. The data that are gathered should be reliable and objective to help you determine if the process is meeting the criteria established by senior management. It provides information to managers at all levels so they can determine actual environmental performance of their existing processes over time to the organization's proposed performance. The basic steps you can follow include:

- Gather the raw data by line employees (those close to the process)
- Analyze the data to make sure it is to be usable
- Place data into a usable format such as a spreadsheet, pareto chart, histogram, etc.
- Compare actual performance against intended performance
- If you're not reaching your goals, conduct a root cause analysis to determine why
- If you're reaching your goals, identify the strengths as to why
- Communicate the results to all stakeholders
- Finally, adjust the processes as necessary

Measuring your day-to-day processes and using that information to improve your procedures is critical to developing a continuous improvement mindset within your organization. Remember the old management saying, "what gets measured gets improved"; EPE is simply an application of that wise axiom.

DEVELOPING YOUR ENVIRONMENTAL PERFORMANCE MEASURES

This discussion will be useful as you develop your outcome measures and your "measurements along the way" within your processes. Keep in mind that measurement is a means to an end, a means to continuous improvement. Your EPEs are measurements made over time, which communicates vital information about the quality of a process, activity, or resource. Consider these thoughts when developing your measurement:

- It "drives" appropriate action
- It shows a trend
- It is unambiguously defined
- It is accepted as meaningful to the customer
- It relates the process to a goal, or the policy
- It is simple, understandable, logical, and verifiable
- It is economical to collect the data
- It is timely, timely, timely

Environmental performance measures are not:

- To be used for disciplinary actions
- One-time snapshots
- Just schedules or charts

Conceptually there can be two different types of EPEs, outcome-oriented measures (QPMs) and process-oriented metrics.

Outcome-oriented measurements appear to be the most common. They are usually a quantitative description of a level of performance or an indicator of performance. They are usually measured relative to a goal or objective. They can also be thought of as the act or process of quantitatively comparing results to requirements to arrive at a quantitative estimate of performance.

Process-oriented measures are taken over a period of time and communicate vital information about a particular process or activity. They show relative performance to control limits defined by the process operation. These target specific information and drive appropriate improvement actions. These specific measurements are typically conducted by line employees as they are the closest to the process. Process measures are more meaningful measures that present data for the specific purpose of waste reduction or process improvement. They focus on:

- Customer requirements
- The input material (hazard, quality, packaging, etc.)
- The process (improving performance to eliminate waste)
- The output (reducing waste, toxicity)
- Contribute to meeting goals, objectives to reduce impact on the environment

As mentioned earlier, a measure must present us with data you can use. It should show a status over time. Remember, only trend data have the potential to be evaluated to the degree needed in order to establish actions to be taken. Before you jump in and start measuring, there are a few criteria to consider in creating a new measure. They must:

- Be meaningful
- Be output-oriented
- Be customer-driven
- Crystallize the environmental policy
- Be quantifiable
- Be simple, understandable, logical, and verifiable
- Show a trend
- Be easily defined and understandable
- Be clearly communicated throughout the organization
- Be timely
- Be regularly reviewed
- Relate process to goal
- Drive the "appropriate action"

First, determine the objective you want to meet, then determine the impact to the environment you want to reduce or minimize. Also, find out what's important to your customer and focus and measure these items; then, if necessary, determine better methods of internal controls and accounting. To focus your efforts measure things like;

- Cycle time
- Defect rate
- Customer satisfaction
- Waste per unit outcome
- Variability

The following are some more questions for coming up with measures for your processes. You can put the measure in the form of a question when you're developing it. This will help everyone to know exactly what the metric is being used for and what it provides.

- *Cycle Time.* If cycle time is the improvement, then what objects or events times are we concerned with tracking? Which part of the cycle time should we be concerned with?
- *On-Time Rate.* If the on-time rate is the improvement, then what object's on-time rate will be improved?
- *Quantity.* If quantity will be the improvement, then what object's or event's quantity will improve? What is the unit/event that this object is measured against?
- *Quality.* Will quality be higher or lower? If quality will be the improvement, then what object's quality will improve?
  - What quality do you expect to improve?
  - What do you expect the improvement to be?
  - What is this increase, decrease, change, or improvement measured against?

Once you have established your EPEs or other measures, it's a good idea to have senior management track them monthly or quarterly at the environmental steering group or other leadership board. This way they keep focused on the overall health of the EMS.

AUDITS UNDER ISO 14000

As we discussed in Chapter 2, the ISO 14001 environmental management standard requires you to continuously monitor the performance of your EMS. This is conducted through the use of performance evaluations and a systematic auditing approach in which you compare your existing management systems against the ISO 14001 management system standard. In short, by conducting an audit, you determine your conformance with the standard.

In general terms, those organizations that have a great commitment to establishing an effective EMS and carefully plan and execute their implementation can expect to be ready to have an EMS audit in 18 to 24 months after implementation begins. Most organizations do not hire an external agency to conduct their first audit; they use internal members. The resources necessary to plan and execute an effective EMS audit are extensive, and numerous consulting firms are available to assist in the planning and implementation of effective EMS development and auditing.

The overall scope and frequency of audits is not clearly defined in the ISO standard nor is it directed to be either an internal (personnel within your organization) or external (personnel outside of your organization) audit. The standard simply directs the audit to be carried out by properly trained and impartial auditors. The standard also discusses that the EMS audit should be based on the relative importance or impact on the environment as well as any findings from previous audits.

As I see it, a typical audit begins with a detailed planning stage, then a document review of the EMS manual and other such material, followed by the auditors visiting the operations and interviewing employees to validate the information found in the document review. These steps will occur whether it's an internal or an external audit.

It is important to note a slight difference in language between the ISO 14001 EMS language and the environmental audit language under the ISO 14010 audit standard. This has to do with the difference between conducting an internal audit simply for senior management information or conducting an external audit to achieve EMS certification.

Although the actual steps you follow are essentially the same, an audit conducted against the ISO 14001 standard is typically a conformity assessment. This occurs when an external (to the organization) auditor, that is employed by a registrar, compares your implemented EMS to the ISO 14001 standard to determine if your EMS conforms to the standard. If it does, then you'll receive a certificate of registration.

Under the ISO 14010 audit, the audit process is the same as the conformity assessment, but the results are presented to senior management so they can understand the health of the EMS. Under the conformity assessment or certification audit the results are presented to a registrar to determine conformance to the standard. Essentially, the organization becomes ISO 14001 certified.

The environmental auditing standard is actually divided into three standards; ISO 14010: General Principles of Environmental Auditing; ISO 14011/1: Auditing Procedures; and ISO 14012: Qualification Criteria for Auditors.[5] These standards are readily available from ANSI, as discussed in Chapter 12. But let's briefly review them before discussing auditing execution (Figure 10.3).

ISO 14010[6]

ISO 14010 is the Guidelines for Environmental Auditing — General Principles of Environmental Auditing. This section is a list of definitions and instructions for auditors and the audit process. It also includes planning suggestions and resource requirements. The overriding theme of this standard is that the audit is checking conformance to a standard; it's not a compliance assessment.

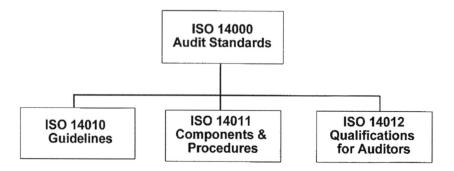

FIGURE 10.3   The three ISO 14000 audit standards.

ISO 14011/1

ISO 14011/1 discusses the procedures to follow during an audit. It includes such things as: definitions, general planning information, determining the audit scope, preparing for the audit, executing the audit, and the audit report.

ISO 14012

ISO 14012 discusses the qualifications and criteria for environmental auditors. The standard outlines the education and work experience and training required for an auditor, in general, and a lead auditor in particular.

AUDITING UNDER EMAS[7]

The auditing requirements under EMAS are similar to those under ISO 14000. Your environmental practices and performance are checked against your stated policy, goals, and regulations. The audit is conducted first to investigate the actions your organization has conducted to improve its processes; then you also audit your management system and structure to assess its effectiveness. Successful programs following the EMAS concept have both regular annual internal audits plus regular continuous improvement emphasis. Periodic internal audits give the necessary information on the health of your processes to make needed improvements. Audits must be conducted every 3 years, and more frequently, depending on the risk of the activities. The EMAS also has a validation step in which an independent accredited environmental verifier reviews the EMAS.

EXECUTING THE AUDIT

Before you consider auditing your EMS, you need to complete virtually all of the steps I have discussed within this text and have a relatively mature EMS in place. The three things you must have prior to considering an audit include:

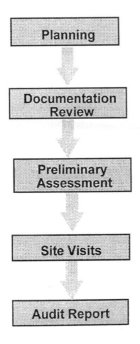

FIGURE 10.4   The EMS audit execution steps.

- A fully implemented EMS
- An EMS manual. This manual outlines the policy, goals, objectives, etc., as well as your commitment to P2 initiatives. This manual serves as the primary document used during the audit
- All personnel who have the potential to impact the environment should have completed training in environmental matters

The steps you will follow to complete an EMS audit, whether you're following EMAS or ISO 14000, trying to achieving certification, or conducting your first internal audit, are virtually the same, and are shown in Figure 10.4 and are discussed in detail below.

DETERMINING THE AUDIT SCOPE

The scope of the audit should be defined from the onset. It should be developed by the P2/ISO 14000 working group and blessed by the environmental steering group. The scope should define the roles and responsibilities of the audit team that essentially becomes the audit team charter. The scope should also cover the manner in which the findings will be reported. You must realize as you develop the audit scope that the resources committed to the audit should be sufficient to meet the scope. In other words, make sure you have the resources necessary to carry out the audit before finalizing the scope. I also recommend that you conduct a preliminary document review while scoping your first audit. This review would include the policy statements, plans, manuals, and records. If the documentation is not complete or is otherwise inadequate, you should inform senior management, identify the inadequacy, and suspend the audit until the documentation is complete or is otherwise adequate.

SELECTING THE AUDIT TEAM

The team membership is very dependent on the overall scope of the audit because the team must include members that have direct knowledge of the processes being audited. Obviously, selecting the audit team is more important for an internal audit. The internal audit team membership should include line employees, shop supervisors, managers, and environmental staff members. Like any other team sponsored by management, a charter should be written for the team and approved by the environmental steering group that outlines the overall responsibilities, authorities, and expected outcome.

PREPARING FOR THE AUDIT

Once the team is established, they should develop an overall execution plan. The ISO 14011 lists specific items you should consider as you develop your audit plan. The plan needs to be detailed, yet flexible enough to permit changes based on information you'll gather during the audit. The audit plan should be presented to and approved by the environmental steering group. As you gather senior management approval, you will probably encounter some objections to the audit plan. If this happens, they should be resolved before the audit begins; this step should be conducted for both internal and external audits. Tools that you can use to help you plan and schedule your audit include gannt charts, pert charts, and other project scheduler tools. These are very helpful in pictorially demonstrating the steps of the audit.

CONDUCTING THE AUDIT

I recommend that the audit begin with an opening meeting where the members of the audit team are introduced to senior management and the scope, procedures, and timetable of the audit are discussed. You may consider developing a team charter and other team building techniques as we discussed in Chapter 3. Once senior management blesses the audit plan, then a document examination phase occurs, followed by interviews and shop visits. The auditors should collect enough evidence to verify conformance to the EMS standard. Remember, the audit will essentially be an in-depth, rigorous, review of your organizational policy, procedures, systems, processes, procedures, measures, and records. Once all of the documentation has been thoroughly reviewed, then on-site visits of systems will occur to validate the data found during the document review. This step is very much like conducting the vulnerability assessment and baseline assessments. Findings, supported by evidence, are then reviewed with the auditor and the organizations audit manager to acknowledge each finding.

The last step is a closing meeting with all responsible parties. At this meeting, the findings are presented to ensure that the parties understand and acknowledge the nonconformance. Any disagreements should be resolved prior to issuing an audit report.

Reviewing Your Documentation

If you're conducting an internal audit, then you'll want to begin with a documentation review once your team is on board and functioning. The purpose is to ensure that the information meets the requirements of the standard. Of course, a full audit will include an on-site assessment, but beginning with the documentation review is critical to a successful audit and could save you unnecessary expenses if your documentation is incomplete.

Conducting a Preliminary Assessment

This is an optional step in which you essentially check the status of your EMS before investing a great deal more effort. Once you have done the records review you have a basic understanding of how well you're meeting your stated EMS goals and objectives. If you are not near target, why bother investing more time and energy? Identify the weaknesses and develop the corrective strategies. If your records review indicated you are on target, then this step can be used to conduct the essential planning for the audit, such as setting up appointments, gathering additional documents, diagrams, and maps and reviewing these.

Site Assessment

The full assessment is conducted on site. It is a rigorous review of your processes typically conducted by 2 to 4 auditors for about a week. It may include interviews, spot checks, and other process overviews, such as those conducted during compliance audits and Malcom Baldridge quality assessments.

Closing Meeting

A closing meeting should be held prior to developing the audit report. The meeting should be a formal presentation to the environmental steering group members on the strengths and weaknesses identified during the audit. At this point you should also discuss the audit report and other communication mechanisms.

AUDIT REPORTS AND RECORDS

The audit report should be prepared under the direction of the lead auditor in a "user-friendly" format so corrective actions and communication can take place after the audit is complete. I recommend that each audit finding or area for improvement should identify which objective is not going to be met or other impact on the policy of the organization. You should also appoint an office of primary responsibility for each area for improvement or audit finding.

ISO 14011 lists specific items to be addressed in the audit report, although the topics could be changed by mutual agreement when you scope the audit. The distribution of the report is done according to the audit plan. Audit reports are the sole property of the client, and the auditors should respect and safeguard confidentiality. Documents pertaining to the audit are retained by mutual agreement. The audit is completed once all the activities in the audit agreement and audit plans are completed.

TAKING CORRECTIVE ACTION

Taking corrective action is the very same step as "Act" in the Plan-Do-Check-Act Cycle; Step 6 (Review and Improve) of the EMS Implementation process shown in Figure 10.1; and Step 8 (Standardizing the Solution) in the P2 Cycle shown in Figure 10.2 . It's simply a matter of fixing all of the problems you identified during the audit. This is easy to say, but probably not that easy to do. The first thing you need to do is prioritize the nonconformance. If you're wondering how you prioritize the discrepancies, then I refer you to Chapter 5 on Vulnerability Analysis. You must prioritize your audit findings based on your EMS goals and objectives. Then as you begin to correct the deficiencies creating the nonconformance, you must fully analyze and understand the root cause of the nonconformance. The methodologies laid out in the Opportunity Assessment phase would be helpful to review as well because the same basic steps will be followed.

CERTIFICATION OF YOUR EMS

As we discussed earlier, an audit of your EMS against the ISO 14001 Standard is typically called a conformity assessment. This occurs when an external (to the organization) auditor, that is, one employed by a registrar, compares your implemented EMS to the ISO 14001 standard to determine if your EMS conforms to the standard. If it does, then you'll receive a certificate of registration. The certification process is an assessment of your EMS against the ISO 14001 standards conducted by an external agency typically referred to as a "registrar". Typically, the registrar is an external firm that specializes in EMS audits. I'd like to further clarify the difference between certification and registration when referring to EMS audits.

- Certification is a procedure followed by a third party who gives written assurance that the product, process, or service conforms to specified requirements. ISO 14001 certification indicates the organization performs in accordance to its EMS.
- Registration is a procedure in which a body indicates relevant characteristics of a product, process, or service, and then includes or "registers" that product, process, or service on a list that is made available to the public.

It's interesting that within the United States we tend to use the term certification interchangeably with registration. Within Europe, EMS certification is the proper term. Confusion can arise, so be careful not to use "registration," if in fact, you mean "certification."[8] Once your organization completes the conformity assessment or certification audit there are three possible outcomes:

- *Approval or certification.* This will occur if your organizations EMS has met all requirements under ISO 14001. This includes both great documentation and full implementation.
- *Conditional certification.* This will occur if your EMS is complete and well documented, but you have not fully implemented it throughout your organization.
- *Disapproval.* Disapproval will occur when documentation or implementation is not complete.

Once you are certified, your work is not over. Typically you will be subjected to periodic surveillance by the registrar. They will re-look at your EMS for conformance to the standards. Typically, once an organization is certified, it is placed on a list or directory published by the certifying body. Thus, these organizations are typically referred to as being ISO 14001 registered. Certification will probably be for a 3-year period, after which another full audit will be conducted. Some certification agencies may offer indefinite certificates but these typically have detailed surveillance requirements.

CLOSING THOUGHTS

The environmental compliance audit has been a very successful tool in maintaining compliance with the numerous environmental laws and regulations, but it was never intended to check on management systems. This is where environmental performance evaluations and EMS audits come into play. These management tools will provide you with the necessary information to continually improve your process and eventually receive certification.

SELF-ASSESSMENT QUESTIONS[9]

These questions have been developed to help you conduct a "table top" audit or self-inspection of your environmental management system and P2 program. Although this isn't entirely inclusive, it provides an outline for you to assess your program's health. Under each statement you will find a scale from 1 to 5. Choose the one value that best describes your organization. The statements are very general descriptions of issues from which a very broad assessment of the health of your EMS can be made. You should apply these questions to your program to help determine where to apply your efforts in the future and evaluate your environmental health over time. A trend analysis of how you rate your organization over time should be an effective way to track your successes.

ENVIRONMENTAL POLICY

1. How well does your organization describe, identify, or articulate its overall environmental policies?

    (POOR) 1   2   3   4   5   (EXCELLENT)

2. Your organization has developed policies that describe the environmental program and its relation to the agency's functions.

    (POOR) 1   2   3   4   5   (EXCELLENT)

3. Your environmental policy considers the nature, scale, and environmental impacts of your activities.

    (POOR) 1   2   3   4   5   (EXCELLENT)

4. Your environmental policy includes a commitment to continual improvement and pollution prevention.

    (POOR) 1   2   3   4   5   (EXCELLENT)

5. Your environmental policy includes a commitment to comply with applicable environmental regulations.

    (POOR) 1   2   3   4   5   (EXCELLENT)

6. Your environmental policy provides a framework for setting and reviewing environmental objectives and targets.

    (POOR) 1   2   3   4   5   (EXCELLENT)

7. Is the environmental policy documented and implemented?

    (POOR) 1   2   3   4   5   (EXCELLENT)

8. The policy is maintained and communicated to all employees.

    (POOR) 1   2   3   4   5   (EXCELLENT)

9. The policy is available to the public.

    (POOR) 1   2   3   4   5   (EXCELLENT)

10. Your organization strives to facilitate a culture of environmental stewardship.

    (POOR) 1   2   3   4   5   (EXCELLENT)

11. Your organization implements programs that aggressively identify and address potential problem areas and emphasize prevention rather than compliance management.

(POOR) 1 2 3 4 5 (EXCELLENT)

12. Your organization has developed a program to address pollution prevention and resource conservation issues.

(POOR) 1 2 3 4 5 (EXCELLENT)

13. Your organization institutes support programs to ensure compliance with environmental regulations and encourages setting goals beyond compliance.

(POOR) 1 2 3 4 5 (EXCELLENT)

14. Your organization integrates the environmental program throughout its operations.

(POOR) 1 2 3 4 5 (EXCELLENT)

15. Concerning the environment, my CEO is conscious of the need to address environmental issues proactively.

(POOR) 1 2 3 4 5 (EXCELLENT)

16. Concerning the environment, my senior management is fully committed, with a main senior manager responsible for and active in environmental matters.

(POOR) 1 2 3 4 5 (EXCELLENT)

PLANNING

Environmental Aspects

1. In assessing your current position on the environment, you have performed structured reviews of all production aspects and determined impacts.

(POOR) 1 2 3 4 5 (EXCELLENT)

2. In assessing your current position on the environment, you have performed structured reviews of aspects and determined impacts of all activities/products/services.

(POOR) 1 2 3 4 5 (EXCELLENT)

3. In assessing your current position on the environment, you have under-taken structured reviews of all aspects and determined impacts of your environmental management systems and practices.

(POOR) 1   2   3   4   5   (EXCELLENT)

4. You have procedures in place which identify those activities that can interact with the environment in order to determine which aspects have significant environmental impacts.

(POOR) 1   2   3   4   5   (EXCELLENT)

5. Are significant impacts considered when setting environmental objectives?

(POOR) 1   2   3   4   5   (EXCELLENT)

6. Is information pertaining to significant aspects kept up to date?

(POOR) 1   2   3   4   5   (EXCELLENT)

Legal Requirements

1. Have you established a procedure to identify and keep current environmental legal requirements affecting your organization?

(POOR) 1   2   3   4   5   (EXCELLENT)

2. Does everybody have access to the legal requirements?

(POOR) 1   2   3   4   5   (EXCELLENT)

3. You organization is aware of and familiar with all environmental regulations which impact the company.

(POOR) 1   2   3   4   5   (EXCELLENT)

4. Your organization is within compliance with all environmental regulations.

(POOR) 1   2   3   4   5   (EXCELLENT)

5. Do you have regular internal audits and measurement systems to ensure that you stay in compliance?

(POOR) 1   2   3   4   5   (EXCELLENT)

6. With regard to verification of compliance requirements we have documented the information needs, procedures, acceptance criteria, the action to be taken, and implemented systems to assess and document the validity of verification information when systems are found to be malfunctioning.

(POOR) 1   2   3   4   5   (EXCELLENT)

Objectives and Targets

1. Your organization has developed the appropriate support structure and developed a strategic plan or environmental management systems to meet identified policy, goals, and objectives.

(POOR) 1   2   3   4   5   (EXCELLENT)

2. How well does your organization develop the strategic plans and management systems to meet the identified objectives?

(POOR) 1   2   3   4   5   (EXCELLENT)

3. Have qualitative objectives and targets been established at each relevant function within the organization?

(POOR) 1   2   3   4   5   (EXCELLENT)

4. Have qualitative objectives and targets been established for all main impacts?

(POOR) 1   2   3   4   5   (EXCELLENT)

5. Have procedures been put into place to update environmental objectives and targets?

(POOR) 1   2   3   4   5   (EXCELLENT)

6. Were relevant legal and other requirements considered when establishing objectives and targets?

(POOR) 1   2   3   4   5   (EXCELLENT)

7. Were technical, financial, and operational requirements considered when establishing objectives and targets?

(POOR) 1   2   3   4   5   (EXCELLENT)

8. Were views of interested parties considered when establishing objectives and targets?

(POOR) 1   2   3   4   5   (EXCELLENT)

9. Are the objectives and targets consistent with the environmental policy?

(POOR) 1   2   3   4   5   (EXCELLENT)

10. Are the objectives and targets consistent with the commitment to pollution prevention?

(POOR) 1   2   3   4   5   (EXCELLENT)

Environmental Management Program

1. Is there an established environmental management program for achieving environmental objectives and targets?

(POOR) 1   2   3   4   5   (EXCELLENT)

2. Does the environmental management program include assignment of responsibilities for achieving objectives and targets at each relevant function and level within your organization?

(POOR) 1   2   3   4   5   (EXCELLENT)

3. Does the environmental management program include the methods and time frame to achieve the objectives and targets?

(POOR) 1   2   3   4   5   (EXCELLENT)

4. Does the environmental management program address new developments or activities?

(POOR) 1   2   3   4   5   (EXCELLENT)

5. Your organization has a detailed EMS manual covering most of its activities which impact the environment.

(POOR) 1   2   3   4   5   (EXCELLENT)

6. Your organization has a detailed EMS manual covering most of its activities, and its procedures for updating, controlling, distributing, and ensuring the manual's use.

(POOR) 1   2   3   4   5   (EXCELLENT)

IMPLEMENTATION AND OPERATION

1. Your organization has developed and implemented the necessary systems to enable personnel to perform their functions consistent with the organization's environmental policy and objectives.

(POOR) 1   2   3   4   5   (EXCELLENT)

2. Your organization has developed and implemented procedures, standards, systems, programs, and objectives that enhance environmental performance and support achievement of organizational goals.

(POOR) 1   2   3   4   5   (EXCELLENT)

Structure and Responsibility

1. Your organization ensures support for the environmental program by management at all levels and assigns responsibilities for carrying out the activities of the program.

(POOR) 1   2   3   4   5   (EXCELLENT)

2. Your organization ensures that all personnel are assigned the necessary authority, accountability, and responsibility to address environmental performance.

(POOR) 1   2   3   4   5   (EXCELLENT)

3. Are the roles, responsibilities, and authorities for the environmental management system defined, documented, and communicated?

(POOR) 1   2   3   4   5   (EXCELLENT)

4. Are the essential resources, namely, human, financial, material, and technology, provided for implementation of the EMS?

(POOR) 1   2   3   4   5   (EXCELLENT)

5. Has top management appointed a specific management representative(s) with defined roles, responsibilities, and authority for establishing, implementing, and maintaining the EMS?

(POOR) 1   2   3   4   5   (EXCELLENT)

6. Do these representatives report to top management on the performance of the EMS for management review as a basis for continuous improvement of the EMS?

(POOR) 1   2   3   4   5   (EXCELLENT)

7. Your organization ensures that employee performance standards related to the environment are clearly defined.

(POOR) 1   2   3   4   5   (EXCELLENT)

8. Your organization ensures that employee performance reviews include the employee's performance related to the environmental performance standards.

(POOR) 1 2 3 4 5 (EXCELLENT)

9. Your organization ensures that exceptional employee performance related to environmental issues is recognized and rewarded.

(POOR) 1 2 3 4 5 (EXCELLENT)

10. Your organization ensures that employee input is solicited.

(POOR) 1 2 3 4 5 (EXCELLENT)

Training, Awareness, and Competence

1. Environmental training needs in your organization have been assessed and addressed in all departments, and systems are in place to update the training as necessary in the future.

(POOR) 1 2 3 4 5 (EXCELLENT)

2. The organization ensures that all personnel are fully trained to carry out the responsibilities of their positions.

(POOR) 1 2 3 4 5 (EXCELLENT)

3. Your organization develops measures to address and improve employee performance.

(POOR) 1 2 3 4 5 (EXCELLENT)

4. Have training needs been identified and appropriate personnel received necessary training?

(POOR) 1 2 3 4 5 (EXCELLENT)

5. Have procedures been established to make employees aware of the importance of conformance with the environmental policy and requirements of the EMS?

(POOR) 1 2 3 4 5 (EXCELLENT)

6. Have procedures been established to make employees aware of significant impacts of their work activities and environmental benefits of improved personal performance?

(POOR) 1 2 3 4 5 (EXCELLENT)

7. Have procedures been established to make employees aware of their roles and responsibilities in achieving conformance with the environmental policy and requirements of the EMS, including emergency response?

(POOR) 1   2   3   4   5   (EXCELLENT)

8. Have procedures been established to make employees aware of the consequences of nonadherance to operating procedures?

(POOR) 1   2   3   4   5   (EXCELLENT)

9. Are personnel who perform tasks that may cause significant environmental impacts competent to perform their duties, based on education, training, or experience?

(POOR) 1   2   3   4   5   (EXCELLENT)

Communication

1. Your organization develops and implements systems that encourage efficient management of information, communication, and documentation.

(POOR) 1   2   3   4   5   (EXCELLENT)

2. Have procedures been established for internal communication about significant environmental aspects and the EMS?

(POOR) 1   2   3   4   5   (EXCELLENT)

3. Have procedures been established for receiving, documenting, and responding to relevant communication from external interested parties as it relates to significant environmental aspects and the EMS?

(POOR) 1   2   3   4   5   (EXCELLENT)

4. Our communications on environmental matters makes sure everybody is aware of the potential environmental effects of their work and of their responsibilities.

(POOR) 1   2   3   4   5   (EXCELLENT)

5. Our communications on environmental matters are carried out effectively through well-established systems, to ensure that everybody is aware of the importance of complying with organizational policy and objectives, of the potential environmental impacts of their work, of their responsibilities, and of the importance of agreed-upon working procedures.

(POOR) 1   2   3   4   5   (EXCELLENT)

6. We have established open communication channels with regulators and respond as necessary to other queries.

   (POOR) 1   2   3   4   5   (EXCELLENT)

7. We have established open communication channels with most interested parties (regulators, customers, suppliers, community members, etc.).

   (POOR) 1   2   3   4   5   (EXCELLENT)

8. We have established open documented communication channels of proven effectiveness with all interested parties and stakeholders.

   (POOR) 1   2   3   4   5   (EXCELLENT)

EMS Documentation

1. Has information describing the core elements of the EMS been documented?

   (POOR) 1   2   3   4   5   (EXCELLENT)

2. Has information been provided which gives direction on where to obtain more detailed information on the operation of specific parts of the EMS?

   (POOR) 1   2   3   4   5   (EXCELLENT)

Document Control

1. Have procedures been established for controlling all EMS-related documents?

   (POOR) 1   2   3   4   5   (EXCELLENT)

2. Are these procedures periodically reviewed and revised, and approved as necessary?

   (POOR) 1   2   3   4   5   (EXCELLENT)

3. Are current versions of relevant documents available at locations where operations essential to the effective functioning of the EMS are performed?

   (POOR) 1   2   3   4   5   (EXCELLENT)

4. Are obsolete documents retained for legal or knowledge preservation purposes suitably identified?

   (POOR) 1   2   3   4   5   (EXCELLENT)

5. Are documents legible, dated (with revisions), readily identifiable, maintained in an orderly manner, and retained for a specified period?

(POOR) 1   2   3   4   5   (EXCELLENT)

6. Are procedures established for creation and modification of various types of documents?

(POOR) 1   2   3   4   5   (EXCELLENT)

7. Our formal environmental documents are comprehensive, reviewed/revised periodically, approved for distribution by authorized personnel, distributed to agreed lists, available in all appropriate sites, and properly removed when obsolete.

(POOR) 1   2   3   4   5   (EXCELLENT)

Operational Control

1. Have the operations and activities that are associated with the significant environmental aspects been identified?

(POOR) 1   2   3   4   5   (EXCELLENT)

2. Responsibilities for operational control and monitoring activities relevant to our environmental performance are fully defined and documented and coordinated across the organization.

(POOR) 1   2   3   4   5   (EXCELLENT)

3. Activities, functions, and processes that affect the environment are known and subject to written work instructions for in-house work.

(POOR) 1   2   3   4   5   (EXCELLENT)

4. Activities, functions, and processes that affect the environment are known and subject to written work instructions for in-house work, monitoring, procurement and contracted work, and approval of planned processes and equipment.

(POOR) 1   2   3   4   5   (EXCELLENT)

5. Have documented procedures been established to cover situations where their absence could lead to deviations from the environmental policy and your objectives and targets?

(POOR) 1   2   3   4   5   (EXCELLENT)

6. Do the procedures stipulate operating criteria?

     (POOR) 1 2 3 4 5 (EXCELLENT)

7. Have procedures related to significant environmental aspects of goods and services from suppliers been established and communicated to suppliers?

     (POOR) 1 2 3 4 5 (EXCELLENT)

8. Have procedures related to significant environmental aspects of contract operations and activities been established and communicated to contractors?

     (POOR) 1 2 3 4 5 (EXCELLENT)

Emergency Preparedness and Response

1. Your organization has developed and implemented a program to address contingency planning and emergency response situations.

     (POOR) 1 2 3 4 5 (EXCELLENT)

2. Your organization has developed and implemented a program to address contingency planning and emergency response situations.

     (POOR) 1 2 3 4 5 (EXCELLENT)

3. Have procedures been established to identify potential for and response to accidents and emergency situations?

     (POOR) 1 2 3 4 5 (EXCELLENT)

4. Have procedures been established for preventing and mitigation of environmental impacts that may be associated with accidents and emergency situations?

     (POOR) 1 2 3 4 5 (EXCELLENT)

5. Do you review and revise your emergency preparedness and response procedures, especially after the occurrence of an accident or emergency situation?

     (POOR) 1 2 3 4 5 (EXCELLENT)

6. Are emergency preparedness and response procedures periodically tested?

     (POOR) 1 2 3 4 5 (EXCELLENT)

Checking and Corrective Action

Monitoring and Measurement

1. Are procedures established to monitor and measure the key characteristics of your operations and activities that can have a significant impact on the environment ?

   (POOR) 1 2 3 4 5 (EXCELLENT)

2. Do the procedures include recording information to track performance, relevant operations controls, and conformance with objectives and targets?

   (POOR) 1 2 3 4 5 (EXCELLENT)

3. Is monitoring equipment calibrated and maintained and a record of the calibration process retained according to your procedures?

   (POOR) 1 2 3 4 5 (EXCELLENT)

4. Do you have a procedure to periodically evaluate compliance with relevant environmental legislation and regulation?

   (POOR) 1 2 3 4 5 (EXCELLENT)

5. Your organization has developed and implemented a program to evaluate progress toward environmental goals and the efficiency of management, support systems, and capabilities. You use these evaluation results and other information to correct deficiencies and improve performance in all facets of operations.

   (POOR) 1 2 3 4 5 (EXCELLENT)

6. Your organization developed a program to assess environmental performance and analyze information resulting from those evaluations to identify areas in which performance is or is likely to become substandard.

   (POOR) 1 2 3 4 5 (EXCELLENT)

7. Your organization has instituted a program of both regular and periodic evaluation performance against stated objectives and has developed procedures to process the resulting information.

   (POOR) 1 2 3 4 5 (EXCELLENT)

8. Your organization instituted a formal program to compare its operations with other organizations, where appropriate.

   (POOR) 1 2 3 4 5 (EXCELLENT)

9. In your organization, environmental monitoring and verification activities are carried out by departments as they see fit, from their resources.

(POOR) 1   2   3   4   5   (EXCELLENT)

10. In your organization, environmental monitoring and verification activities are defined and documented by most departments.

(POOR) 1   2   3   4   5   (EXCELLENT)

11. In your organization, environmental monitoring and verification activities are defined and documented and fully resourced in all departments.

(POOR) 1   2   3   4   5   (EXCELLENT)

Nonconformance and Corrective and Preventive Action

1. Your organization implemented an approach toward continuous improvement that includes preventive and corrective actions, as well as searching out new opportunities for programmatic improvements.

(POOR) 1   2   3   4   5   (EXCELLENT)

2. Are procedures established for investigating nonconformance with the environmental policy and EMS and for taking action to mitigate the impacts caused by nonconformance?

(POOR) 1   2   3   4   5   (EXCELLENT)

3. Are the corrective or preventive actions taken to eliminate the causes of nonconformance appropriate to the magnitude of the problem and the environmental impact?

(POOR) 1   2   3   4   5   (EXCELLENT)

4. With regard to investigation and corrective action, responsibility is defined and procedures are in place to investigate, plan, take action, assess effectiveness, and change procedures as a result.

(POOR) 1   2   3   4   5   (EXCELLENT)

Records

1. Are procedures established for identification, maintenance, and disposition of environmental records (includes training records and results of audits and internal reviews)?

(POOR) 1   2   3   4   5   (EXCELLENT)

2. Are environmental records legible, identifiable, and traceable to operations or activities?

(POOR) 1 2 3 4 5 (EXCELLENT)

3. Are these records readily retrievable and protected against damage, deterioration, or loss?

(POOR) 1 2 3 4 5 (EXCELLENT)

4. Is a specific retention time established for these records?

(POOR) 1 2 3 4 5 (EXCELLENT)

5. Our environmental records are kept according to a defined system covering storage, maintenance, and retention times for all environmental management activities; and covering all objectives and targets addressing procurement and contracted work, with established policies on internal and external availability.

(POOR) 1 2 3 4 5 (EXCELLENT)

Environmental Management System Audit

1. Have a program and procedures been established for periodic environmental management system audits?

(POOR) 1 2 3 4 5 (EXCELLENT)

2. Does the EMS audit conform to the requirements of the ISO 14001 standards?

(POOR) 1 2 3 4 5 (EXCELLENT)

3. Are the audit results presented to management for review?

(POOR) 1 2 3 4 5 (EXCELLENT)

4. Regarding the implementation of our EMS or environmental performance, we have a defined plan and protocol for independent, internal or external auditing of each area covering both environmental effects and management systems.

(POOR) 1 2 3 4 5 (EXCELLENT)

MANAGEMENT REVIEW

1. How well does your organization review environmental policies, programs, and support structures?

   (POOR) 1   2   3   4   5   (EXCELLENT)

2. Your organization reviews your environmental performance, policy, and goals on a regular basis.

   (POOR) 1   2   3   4   5   (EXCELLENT)

3. Does the top management periodically review the EMS to ensure its continuing suitability, adequacy, and effectiveness?

   (POOR) 1   2   3   4   5   (EXCELLENT)

4. Does the management review address the possible need for changes to policy, objectives, and other elements of the EMS, in light of the EMS audit results?

   (POOR) 1   2   3   4   5   (EXCELLENT)

5. Do you take steps to "fine tune" your program based on feedback from customers and suppliers?

   (POOR) 1   2   3   4   5   (EXCELLENT)

POLLUTION PREVENTION

1. Have you adequately planned to execute an effective pollution prevention program?

   (POOR) 1   2   3   4   5   (EXCELLENT)

2. Does your EMS address the appropriate key focus areas of hazardous waste, solid waste reduction, and resource conservation?

   (POOR) 1   2   3   4   5   (EXCELLENT)

3. Does your organization have a 5-year plan or program to implement its P2 efforts? Have you articulated this plan or program in your budget submittals?

   (POOR) 1   2   3   4   5   (EXCELLENT)

4. Is your organization eliminating ozone-depleting chemicals from all facility utility systems?

   (POOR) 1   2   3   4   5   (EXCELLENT)

5. Does your organization have a municipal solid waste (MSW) reduction program? Are you tracking total waste disposal in order to ensure that solid waste reduction goals are met?

   (POOR) 1   2   3   4   5   (EXCELLENT)

6. Are you preventing pollution in every area throughout the organization?

   (POOR) 1   2   3   4   5   (EXCELLENT)

7. Does your organization operate a hazardous material reuse/reissue center? Do you track hazardous material purchases?

   (POOR) 1   2   3   4   5   (EXCELLENT)

8. Does your organization have an established affirmative procurement program for all designated EPA guideline items purchased?

   (POOR) 1   2   3   4   5   (EXCELLENT)

9. Do you have a meaningful way to reevaluate your progress on a regular basis?

   (POOR) 1   2   3   4   5   (EXCELLENT)

10. Do you have and are you tracking measures for (grade each one):

    (POOR) 1   2   3   4   5   (EXCELLENT)

    - Hazardous waste reduction?
    - EPA 17 reduction?
    - HAZMART implementation?
    - VOC reduction?
    - CFC chiller elimination?
    - ODC use?
    - Solid waste reduction?
    - Recycling rate?
    - Affirmative procurement?
    - Water use?
    - Energy conservation?

11. Have you conducted a pollution prevention baseline assessment?

    (POOR) 1   2   3   4   5   (EXCELLENT)

12. Have you conducted pollution prevention opportunity assessments?

    (POOR) 1   2   3   4   5   (EXCELLENT)

13. Have you implemented P2 opportunity assessment recommendations?

(POOR) 1   2   3   4   5   (EXCELLENT)

14. Does your organization have a plan to conduct a water use prioritization survey and a comprehensive facility audit in the water conservation area?

(POOR) 1   2   3   4   5   (EXCELLENT)

15. Do you have a long term plan to conduct or obtain comprehensive facility audits, based on prioritization surveys performed for both energy and water conservation efforts?

(POOR) 1   2   3   4   5   (EXCELLENT)

16. Do you have a pollution prevention outreach effort to seek out innovative ideas and to share its own success stories?

(POOR) 1   2   3   4   5   (EXCELLENT)

17. Do you have a listing of each ozone-depleting chemical used, its user, and the process it is used for?

(POOR) 1   2   3   4   5   (EXCELLENT)

REFERENCES AND NOTES

1. Voluntary Environmental Self-Policing and Self-Disclosure Interim Policy Statement, *Federal Register,* April 3, 1995, p. IIA.
2. Environmental terms taken from EPA's homepage on the internet, http://www.epa.gov/docs/OCEPAterms/eterms.html, Feb. 17, 1996.
3. Environmental Auditing: A Useful Tool That Can Improve Environmental Performance and Reduce Costs, Report No. GAO/RCED-95-37, Government Accounting Office, Washington D.C., April 1995.
4. Environmental Compliance Assessment and Management Program, U.S. Air Force, Aug. 19, 1988, p. 3.
5. Modified from Cascio, J., *The ISO 14000 Handbook,* ASQC, Milwaukee,1996, p. 305.
6. A lot has been written on the ISO 14000 audit standards. Before you consider conducting an audit I recommend that you search the internet for more information. More information including internet addresses can be found in Chapter 12. For detailed information on the audit standards, you can obtain the standards directly from ANSI. Their address and phone number are also listed in Chapter 12.
7. Adapted from readings in the E.C. Eco-Management and Audit Scheme for UK Local Government, EMAS Help Desk Guidance Notes, December 1995, United Kingdom Department of the Environment.
8. Discussion adapted from Hemenway, C.G., *What Is ISO 14000? Questions and Answers,* ASQC, Milwaukee,1995, p. 32.

9. This series of self assessment questions came from a number of different sources including: a. A draft EPA document, Code of Environmental Management Principles (CEMP) for Federal Agencies, undated; b. A questionnaire developed by a NATO working group to review the status of environmental management programs throughout NATO countries. A copy was provided by Col. Rick Drawbaugh from the office of the Undersecretary of the Air Force for Environment, Safety and Health (SAF/MIQ); c. A questionnaire developed by my colleague Dr. Bill Handcuff, a private consultant specializing in pollution prevention and environmental management systems. With permission.

11 The Pre-Visit Questionnaire

As I discussed in the earlier chapters, it is necessary to send out a pre-visit questionnaire to those shops or other groups that you plan on visiting to conduct baseline assessments or opportunity assessments or perhaps even environmental audits. The information in the chapters on vulnerability assessment, baselines and opportunities assessment, and the chapter on audits, along with the information in this chapter will help you develop a very effective questionnaire for your organization.[1] Much of the information in this series of questionaires may be available from record searches. I recommend that you use that information first to answer as many of the questions as possible before sending them out to various shops. Then have the shops validate the information from the records review. If no information is available, then provide the questionnaire to the shop and assist them as necessary in completing the information.

QUESTIONNAIRE INTRODUCTION

The purpose of this questionnaire is to collect information for a pollution prevention assessment. The objective of this assessment is to determine the baseline of how much and what types of waste are generated from your facility or shop. Once the types and amounts of wastes are understood and identified, then subsequent steps to eliminate the waste can take place.

Information to complete this questionaire should be for the baseline year. Wherever possible, use records to provide data. Estimates are acceptable; however, please note where estimates are used. Please complete all applicable questions.

A follow-up visit will be made by a baseline survey team to discuss and clarify information requested on the questionnaire and to interview you and other shop members in an effort to fully understand your process. If you're responsible for more than one shop, please copy and complete a questionnaire for each shop.

QUESTIONNAIRE

FACILITIES INFORMATION

1. Location:
   Facility name (shop name):
   Address:
   Name of person completing form:
   Phone number:
   Alternate contact and phone number:

2. Facilities you are completing this questionnaire for:
3. Briefly describe what your facility does (e.g., "We perform full and touch-up paint jobs on various types of equipment").
4. Briefly describe the specific steps in the activities your facility performs (e.g., for paint shop: "1. Paint is removed from equipment by abrasive blasting; 2. A naphtha wipe-down is performed; 3. Primer is applied; 4. Paint is applied; 5. Decals applied.").
5. What is the measure of output for this process?
6. What are the approximate dimensions of the process or shop area?
7. Provide a sketch of the process and area.
8. Are engineering drawings available? If so, provide a copy or a point of contact.

LABOR REQUIREMENTS

1. How many workers are involved in the process?
2. Is special environmental and/or safety training needed?
3. What are the hours of operations (hours per day, days per week, weeks per year).

EQUIPMENT AND MATERIAL REQUIREMENTS

1. Is the work done manually? If not, describe the system?
2. Identify the major equipment used in the process (e.g., by size, use, manufacturer, efficiencies, controls (temperature, pressure) age, etc., dependent on the process).
3. What are the procedures for cleaning and or maintaining equipment? What material is used, in what concentration, and in what quantity?
4. What are the methods of disposal for the waste material?
5. Are there any other shops that perform similar operations at this facility?
6. Are there any special material specifications for raw materials (weight, color, heat resistance, density).
7. Where are the materials purchased?
8. List contact name and phone number.
9. What is the normal procedure for ordering materials ?
10. How often are materials restocked?
11. Please provide information on raw material usage on Figure 11.1. You do not need to include chemicals with usages less than 1 gallon per year or aerosols (such as spray paint) with less than 6 cans per year usage. Solid materials (such as chemical powders and welding rods) do not need to be included.
12. Please provide information on water used within your processes by completing Figure 11.2
13. Please provide information on energy used within your process by completing Figure 11.3.

Type	Product Name and Manufacturer	Use	Quantity Used/Month	Cost	Size of Supply Containers

FIGURE 11.1   Raw material usage form.

Process Stream	Quantity Used / Month	Use	Cost
Input			
Output			
Recycle			

FIGURE 11.2   Water usage information.

HAZARDOUS WASTE GENERATION

1. Has an analysis of the waste streams been done?
2. If so, can a copy of the analysis be provided?
3. List contact name.
4. List contact's phone number.
5. Sketch of waste storage area (note size and number of containers).
6. What group/department is responsible for picking up the wastes? Provide contact name and phone number.
7. How often are the wastes picked up?
8. What is the procedure for arranging for pickup?
9. Are wastes segregated?
10. Complete the list below (Figure 11.4) concerning hazardous material spilled during the baseline year.

Type (fuel/source)	Quantity used/month	Use	Cost
[electric]			
[gas]			
[oil]			

FIGURE 11.3   Energy usage form.

Name of Spilled Material	Quantity Spilled (gallons/lbs)	Clean-up Method	Total Amount Cleaned-up (gallons/lbs)

FIGURE 11.4   Spilled hazardous material form.

11. Does your facility perform welding? If so, indicate the rod type, pounds of rod used, and type of welding completed.
12. Do you have any degreasers or solvent cleaner tanks? If so, complete Figure 11.5 for EACH degreaser.
13. Oil–water separators: Does your facility/shop have any sources which drain to an oil–water separator? If so, list the source.
14. How often is the oil–water separator checked (oil skimmed, checked for proper operation/condition)?
15. How often is oil removed from the oil–water separator? _____

Name of Solvent In Tank	Tank Dimensions	Tank Volume	Frequency of Solvent Change-out	If tank has a lid, how long is it open a week

FIGURE 11.5 Degreasing and solvent tanks information.

Note: The P2 baseline team should obtain a list of oil–water separators from facilities engineering, maintenance, or public works. Also, obtain a map showing the separator locations and tie-ins into sanitary sewers and stormwater sewers. If you plan to focus on separators as a potential source of pollution, you may also want to obtain sampling information for oil–water separators and stormwater outfalls.

16. Please complete Figure 11.6 on hazardous and nonhazardous industrial wastes.
17. Please complete Figure 11.7 on hazardous waste characterization data.

SOLID WASTE QUESTIONS

Which of the following facilities/programs are utilized for the disposal of non-hazardous solid waste (include name of facility, contact person and phone number)?

1. Solid waste landfill.
2. On-site or off-site? If on-site, is it in operation or closed?
3. Material recovery facility (a facility that separates materials for recovery prior to disposal).
4. Construction/demolition landfill. On-site or off-site? If on-site, is it in operation or closed?
5. Asbestos landfill. On-site or off-site? If on-site, is it in operation or closed?
6. Recycling facility.
7. Transfer station (a facility that accepts materials for transportation to a disposal facility).
8. Composting facility.

Waste Stream	Quantity generated/month	Source or origin	Disposal method	Disposal cost

FIGURE 11.6   Hazardous and nonhazardous industrial waste information.

Waste Stream	Physical form (solid, semi-solid, liquid, etc.)	Major Constituents	
		Chemical Substance	Concentration

FIGURE 11.7   Hazardous and nonhazardous industrial waste characterization form.

9. Incineration (a facility that thermally decomposes material and may or may not include energy recovery). If so, check type of incineration:
   Municipal waste combustor
   Air curtain destructor
   Open burning at landfill
10. Does your facility generate any sludge from wastewater or water treatment processes? If yes, how are the sludges disposed of?
11. Do you generate ash from thermal combustion? If yes, where are ashes generated from and how are the ashes disposed?

12. What are the quantities of waste disposed of in these facilities for the baseline year?
      Landfilled
      Material recovery facility
      Recycling facility
      Transfer station
      Composting facility
      Incineration
      Medical waste
      Nonhazardous solid waste
      Air curtain destructor
      Open burning
      Sludges
      Ash
13. What were the baseline year costs for solid waste collection/transportation, including annual labor, equipment, and operation provided by your organization?
14. What were the baseline year costs for contract collection/transportation, including contract overhead and average tipping fees?
15. Have you conducted a waste stream analysis such as a waste characterization study? If yes, can you provide records on the study? When was study conducted?
16. Is your organization part of a regional or a local solid waste management plan?
17. Do state or local regulations require recycling? If yes, what are the requirements for recycling?
18. If you have recycled materials in the baseline year, where have the recycled materials gone?
19. What are the operational costs for the recycling program?
20. What was the facility population for the baseline year and what are the projections for each year?

CORROSION CONTROL AND PAINTING

1. What items do you paint?
2. What is the typical frequency and reason for performing a full paint job? What is the typical frequency and reason for performing partial paint jobs?
3. Provide a list of all paints and solvents and their usage quantities along with other chemicals on the attached chemical usage form for the baseline year.
4. Provide a list of any lead-based or chromium-based paint that is used.
5. Do you test old paints that are being stripped or removed by abrasive blasting for lead or chromium content?
6. Is a paint booth used?

7. Do you have a waterfall paint booth? If so, how do you dispose of paint booth wastewater (if any) and paint solids/sludge? Include the quantity of sludge disposed of on the attached "Hazardous and Industrial Nonhazardous Waste" form at Figure 11.7.
8. Do you have filters on the paint booth? If so, are they fiberglass, fabric, other?
9. Where are they disposed?
10. Do you use a paint gun to paint?
11. If so, what type of gun is used? (Check all that apply — for more than one indicate percent of total paint sprayed through gun)
    High volume low pressure
    Airless aerosol
    Air gun
    Electrostatic
    Other (specify)
12. Do you do any fiberglass work?
13. If so, do you sand the fiberglass?
14. What type of sander is used?
15. How much dust is disposed of (per week, per month)?
16. Include chemicals used in fiberglass work on chemical usage form.
17. Do you sand with mechanical sanders?
18. If so, where do you sand the equipment (outside, in hangar, in booth)?
19. Does your facility use spray paints? If so, please list on the attached chemical usage form. What do you spray paint and why (e.g., stenciling, touch-up painting for aesthetics, etc.)?
20. Does your facility perform abrasive blasting? If so, complete the following:
    a. Is abrasive blasting performed in an enclosed unit, inside a building (e.g., paint booth), or outdoors?
    b. Is there a dust-control device (baghouse, cyclone, or filters [as in a paint booth])? If so, what type of control(s)?
    c. What type of medium is used?
    d. What is being blasted off? Does it contain lead, chromium, or any other hazardous material?
    e. How many pounds of media are used in a year?
    f. How many pounds of spent media are disposed in a year?

SERVICE STATION AND VEHICLE MAINTENANCE

1. How many service bays do you have?
2. What quantities of fuels did you pump in the baseline year (Figure 11.8)?
3. Do you have stage II vapor recovery on your dispensing nozzles? If so, provide the equipment manufacturer and system efficiency.
4. Do you have a Freon recovery/reclaim unit? If so, provide quantities of Freon recovered in the baseline year and disposition of material recovered.
5. Complete Figure 11.9 for each of your fuel storage tanks.

Type (fuel/source).	Quantity used/month	Number of Nozzles	Remarks
Regular Leaded			
Regular Unleaded			
Gasohol			
Premium			
Diesel			

FIGURE 11.8   Dispensed fuel quantities form.

Tank Number or Name	Capacity and Dimensions	Type of Fuel	Vapor Recovery
Tank 1			
Tank 2			
Tank 3			

FIGURE 11.9   Fuel storage tank information.

6. If your tanks have stage I vapor recovery, indicate which tanks have this, the efficiency of the system, and the equipment manufacturer.
7. What items do you paint?
8. If you perform vehicle painting, what is the typical frequency and reason for performing a full paint job? What is the typical frequency and basis for performing partial paint jobs?
9. Provide a list of all paints and solvents and their usage quantities along with other chemicals on the attached chemical usage form for the baseline year.
10. Provide a list of any lead-based or chromium-based paint that is used.
11. Do you test old paints that are being stripped or removed by abrasive blasting for lead or chromium content?
12. Is a paint booth used? If so, complete questions *a* and *b*.
    a. Do you have a waterfall paint booth? If so, how do you dispose of paint booth wastewater (if any) and paint solid/sludge? Include the quantity of sludge disposed on the attached "Hazardous and Industrial Nonhazardous Waste" form.

    b. Do you have filters on the paint booth? If so, are they fiberglass, fabric, other?
      Filter Dimensions (length × width × depth).
        No. of filters in booth? _____
        How often changed? _____

13. Do you use a paint gun to paint? If so, what type of gun is used (check all that apply — for more than one, indicate percent of total paint sprayed through gun)?
    HVLP (high volume low pressure)
    Airless aerosol
    Air gun
    Electrostatic
    Other (specify)

14. How often do you perform engine fluid changes (i.e., oil, antifreeze, hydraulic fluid, etc.)?

15. What is the basis for performing the changes (i.e., manufacturer recommendation)?

16. How are asbestos clutch disk, brake pads, and shoes disposed of?

17. Are air conditioning services being done? Is a Freon recovery system being used?

18. Are all mechanics servicing air conditioning systems certified and registered with EPA?

19. How much Freon did you use in the baseline year?

VEHICLE EMISSIONS

1. Does your facility use vehicles? If so, list the type of vehicle, fuel used, and number of miles driven for the baseline year (Figure 11.10) (note that this information is important for air emissions).

For type of vehicle, use these descriptions.
- Light-duty gasoline-powered vehicles (LDGV) are gasoline-fueled vehicles designed primarily for transportation of persons and having a capacity of 12 persons or less.
- Light-duty diesel-powered vehicles (LDDV) are diesel-fueled vehicles designed primarily for transportation of persons and having a capacity of 12 persons or less.
- Light-duty gasoline-powered trucks I (LDGTI) are light-duty gasoline-fueled trucks with gross vehicle weight ratings of 6000 pounds or less.
- Light-duty gasoline-powered trucks II (LDGTII) are light-duty gasoline-fueled trucks with gross vehicle weight ratings between 6001 and 8500 pounds. This vehicle classification is required, since these trucks were classified as heavy-duty vehicles through the 1978 model year.
- Heavy-duty gasoline-powered vehicles (HDGV) are gasoline-fueled motor vehicles designed primarily for the transportation of property and rated at more than 8500 pounds gross vehicle weight, or designated

Model Year of Vehicle	Type of Vehicle	Fuel Type	Miles Driven or Hours Operated	Percent Driven at the Plant

FIGURE 11.10   Motor vehicle information.

primarily for transportation of persons and having a capacity of more than 12 persons.

- Light-duty diesel-powered trucks (LDDT) are diesel-fueled trucks designed primarily for transportation of property and rated at 8500 pounds or less gross vehicle weight.
- Heavy-duty diesel-powered vehicles (HDDV) are diesel-fueled motor vehicles designed primarily for the transportation of property and rated at more than 8500 pounds of gross vehicle weight.
- Motorcycles (MC) are motor vehicles designed to travel with no more than three wheels in contact with the ground, and with a curb weight less than 1500 pounds.

Fire Departments

1. Do you have routine fire fighting drills? What fire extinguishing media do you use in these drills (Halon, foam, etc.)? How much was used in the baseline year?
2. What flammable materials do you use to stage fires? How much was used in 1992? How is the unused waste material disposed of, and how much was disposed of?
3. What type of extinguishers do you use? How much of each type of medium did you add to the extinguishers in the baseline year?
4. What is the quantity of Halon, or equivalent fire fighting foam, contained in systems? (attach a list of systems if necessary)?

Hospitals and Health Facilities

1. How much medical waste was disposed of in the baseline year? Please write these quantities on the attached hazardous waste form.
2. How was medical waste disposed of?

3. Does the hospital have an incinerator? If so, how many pounds of medical waste were incinerated in the baseline year?
4. What type of waste was incinerated?
5. Is there a radiology/imaging department that generates wastes such as fixer and developer? Please list these wastes, quantities generated, and whether waste is disposed of or recycled on the hazardous waste form.
6. Does your facility have back-up generators? If so, provide the following information on KW rating and hours operated or amount of fuel consumed for each generator:

FUEL FARMS

1. What types of fuel are received in the area (JP-4, MOGAS, DF-1, DF-2, kerosene, etc.)? How are they delivered to the facility (via pipeline, rail car, tank truck)?
2. How many gallons of each type of fuel were received in the baseline year? Complete Figure 11.8 as necessary.
3. Provide information on:
    Average daily temperature
    Average daily temperature change
    Average wind speed in mph or knots
4. Provide information for each underground storage tank. Include tanks in the reclaimed/waste fuels area and fuel tanks at transfer stations.
    Tank identification no.
    Building/facility near tank
    Horizontal or vertical
    Capacity
    Type of fuel stored
    Throughput for each tank (gallons/year for the baseline year).
    Type of fill: submerged, splash, other (specify)
    Stage I vapor recovery present? If yes, efficiency.
    Average daily liquid temperature
5. Provide the following information for each aboveground storage tank (make copies as needed). Include tanks in the reclaimed/waste fuels area, and fuel tanks at transfer stations.
    Tank identification no.
    Building no./facility near tank
    Horizontal or vertical capacity (gal)
    Dimensions: diameter, height or length
    Type of fuel stored
    Tank color: shell, roof
    Type of roof for aboveground tanks:
        Fixed roof tanks
        External floating roof tanks
        Internal floating roof tanks (or floating pan)
        Variable vapor space tanks

Throughput for tank (gallons/year in baseline year).

Is the tank welded or riveted?

Number of pressure or vacuum valves. Vacuum/pressure setting.

Stage I vapor recovery present? If yes, efficiency.

Average daily liquid temperature.

6. External floating roof tanks (for each tank)

    Tank identification no.

    Type of seal

    Type and number of fittings

7. Internal floating roof tanks (for each tank)

    Tank identification no.

    Type of seal

    Type and number of fittings

8. Provide a description of emissions control equipment (conservation vents, submerged fill lines, carbon adsorption, vapor balance, etc.). Indicate which tanks they are installed on.

9. Provide a description of fuel transfer operations (tank-to-tank, tank-to-refueling vehicle, tank-to-plane or vehicle, refueling vehicle-to-plane). Please provide a flow diagram for each fuel type used, which shows transfers of fuel from receipt to use in equipment.

10. Where are refueling vehicles maintained? Provide a contact name and phone number.

11. Where are refueling vehicles serviced (fuel, oil addition, washed, etc.)? Provide a contact name and phone number.

12. Is fuel from maintenance areas returned to you for re-use or disposal? If so, how much and how did you dispose of it?

BOILERS/FURNACE/HEATING

1. For each boiler/furnace provide the following information. Use the attached form to answer each question. (Make copies as needed)

    a. Firing capacity (BTU/Hr) of the boiler/furnace for each of the following fuels: natural gas, distillate fuel oil, residual fuel oil, and coal.

    b. Fuel used during the baseline year: natural gas ($ft^3 \times 10^6$), distillate fuel oil (gallons), residual fuel oil (gallons), coal (tons). If it is not known for each boiler, please provide fuel usage for each fuel type for groups of boilers.

    c. Indicate the type of boiler if coal is burned: pulverized coal, wet bottom, dry bottom, spreader stoker, overfeed stoker, underfeed stoker, handfired, other.

2. Answer the following:

    a. Do the boilers/furnaces have control equipment for particulate matter? If so, what type and which boilers/furnaces (cyclone with no reinjection of ash, cyclone with reinjection of ash, electrostatic precipitator, baghouse, wet scrubber)?

Location or Boiler Number	Input Capacity BTU/Hr	Fuel Type	Quantity of Fuel Used	Operating Hours	Type of Furnace

FIGURE 11.11   Boiler and furnace information.

    b. Are the boilers/furnaces equipped with any NO_x controls? If so, what type and which boilers/furnaces (low NO_x burner, flue gas recirculation, low excess air [LEA] burner, other)?

    c. Do the boilers/furnaces have control equipment for sulfur dioxide emissions? What type of device and which boilers/furnaces?

3. If distillate or residual fuel oil is used, provide the actual weight percent sulfur content for the fuel (supplier should have this information).

4. If coal is used, provide the actual weight percent sulfur content for the coal (supplier should have this information).

5. If stack tests have been performed recently, provide copies of the stack test reports for those boiler/furnaces tested.

6. Describe any other maintenance performed on the boiler/furnaces.

7. How much ash was generated in the baseline year and how was the ash disposed? Include quantity disposed on the "Hazardous and Industrial Nonhazardous Waste" form, Figure 11.4.

8. Where do you discharge your boiler blowdown?

9. Where do your steam traps discharge? Do you recycle condensate?

10. Complete Figure 11.11 for each boiler or furnace.

CHILLERS/AIR CONDITIONING

1. Provide a list of the air conditioning systems on base and the amount of Freon and type of Freon contained in each air conditioner.

2. How much of each type of Freon was used in the baseline year?

3. Do you "bank" or store Freons?

4. Describe any other maintenance performed on refrigeration equipment, chillers, and associated equipment (cooling towers).

5. Have service personnel received training/certification on how to work with Freon refrigeration systems?

ELECTRICAL

1. Do you handle PCB-containing transformers, ballasts or switches? Is there a program to remove these from service? If so, describe program. Please include disposal information on the "Hazardous and Nonhazardous Industrial Waste" form.
2. Do you handle asbestos containing insulated wire or spark shields? Is there a program in place to deal with disposal of these materials when taken out of service? Please include disposal information on the attached "Hazardous and Nonhazardous Industrial Waste" form.
3. Please include disposal of telephone poles and other excess wire on the "Municipal Solid Waste" Questionnaire.

FACILITY DEMOLITION AND MAINTENANCE

1. Were any buildings demolished/disposed in the baseline year?
    a. If a contractor demolished/disposed of the buildings, consult the contractor for the following information.
    b. What was the typical material of construction?
    c. How much debris was disposed of (in cubic yards)?
    d. If the quantity of material disposed of is unknown, describe the building and indicate the square footage and the material of construction of the buildings disposed of.
2. Do you have particulate emission controls for collecting sawdust from the wood shop (e.g., vacuum ducts at saws connected to a cyclone which dumps the sawdust into a covered hopper outside)?
    a. If yes, what type of control equipment do you have and what is the efficiency? Describe your control equipment.
3. How do you dispose of the saw dust collected and how much did you dispose of in the baseline year? Please enter the quantity disposed on the attached "Municipal Solid Waste" form.
4. Is the shop door generally open or closed when operating wood working equipment?
5. How do you dispose of your scrap wood? Please enter the quantity disposed on the attached "Municipal Solid Waste" form.
6. Do you use a paint gun to paint? If so, what type of gun is used? (Check all that apply — for more than one indicate percent of total paint sprayed through gun).
    High velocity low pressure
    Airless aerosol
    Air gun, electrostatic
    Other (specify)
7. Do you perform traffic painting using a painting machine? If so, please provide the manufacturer and model of the machine and provide the name, and amount of the paint used in the machine.

Engine Description	Engine Type (2 stroke, 4 Stroke)	Engine Size (Hp)	Fuel Type	Hours of Operations/ Year

FIGURE 11.12   Grounds-keeping equipment information.

ENTOMOLOGY AND GROUNDS KEEPING

1. List the names and quantities of pesticides, herbicides, and other chemicals used in the baseline year on the "Raw Material Usage" form.
2. List amounts of expired shelf-life pesticides, herbicides, and other chemicals disposed of on the "Hazardous and Industrial Nonhazardous Waste" form.
3. Complete the attached "Entomology and Lawn Maintenance Equipment" form for fuel-burning equipment used by your facility, including chain saws and other related equipment.
4. Which organization maintains your lawn maintenance equipment?
5. Complete Figure 11.12 on ground keeping equipment.

AIR EMISSIONS

Many types of air emissions have been identified already on this checklist. Use form 11.13 to identify any you may have missed.

1. Has an air permit application been filed?
2. Who is the person responsible and his/her phone number
3. Is a copy available?
4. Complete Figure 11.13.

OTHER POLLUTION PREVENTION INFORMATION

1. Have other P2 initiatives been implemented?
2. What is seen as the main operating constraint?
3. Are there any flow diagrams, specifications, etc. that may be helpful?

Substance	Quantity/month	Source of estimate

FIGURE 11.13   Air emissions table.

REFERENCE AND NOTES

1. This series of questions was adapted from approximately 25 questionnaires I developed with the consulting firm Law Environmental to conduct P2 baseline surveys and opportunity assessments at 34 DoD Installations. I believe this series of questions will be helpful to you as you develop your own survey or questionnaire.

12 Continually Improving Your EMS

After you've established your EMS and you've been on your journey for a while, you'll probably have a plateau in your success. Your future successes may be dependent on more detailed process analysis that will be required to improve more complex processes. There are two tools you can use to help, COPIS and benchmarking. I'll discuss these below and then provide a list of resources where you can obtain more information on pollution prevention and environmental management systems.

COPIS

This is a great tool to use when you're analyzing your processes. You could use this tool or procedure when you first develop your strategic plan, goals, and measures, but it's too easy to get bogged down and lose focus at that point in your journey. It seems more appropriate to apply this procedure after you've had a few successes. The COPIS spreadsheet is shown in Figure 12.1. COPIS stands for: Customers, Outputs, Processes, Inputs, and Suppliers.[1] Essentially, it's a 10-step method that should be completed by the process owners — those at the first level of contact with the customers, suppliers, products, and services. Have the person that provides service, creates the product, and works directly with the customers and suppliers on a particular task complete this document.

I use this tool to really understand processes, not from a technical perspective, but from a managerial perspective prior to attempting to make process improvements. Essentially you go through an analysis of determining who your customers are, defining your customers' requirements, and evaluating what your suppliers are providing you. With this detailed understanding of a particular process, you compare your existing process to what it could be to determine if there is room for improvement. If there is, you can tie this into your EMS manual or strategic plan: your objectives, strategies and measures. COPIS can help you decide which processes to focus on, where you want to be, how fast you will go, how you will get there, and chart out the obstacles in your way. This tool, along with your strategic plan, flowcharts, and audit results helps chart your roadmap to future successes. The basic steps you follow are:

- Review your key result areas and goals
- Review your objectives
- Review your audit results
- Review your existing processes as written in your strategic plan and EMS manual

1. PRODUCTS AND SERVICES	2. RECIPIENTS OF PRODUCTS AND SERVICES/CUSTOMERS	3. KEY CUSTOMER QUALITY REQUIREMENTS/INDICATORS	4. KEY CUSTOMER SATISFACTION INDICATORS	5. VALUE ADDED ACTIVITY
Products are tangible things you give to people: reports, written policy, guidance, and regulations funding, briefings, data, money, funding, resources. Services are things you do for people: Inspections, analysis, advice, validation, verification, processing paper work, coordination.	Your *immediate* customer. Keep in mind that you may be one supplier in a chain of suppliers supporting a customer in the field. You should know your immediate customer requirements and final customer requirements. You should identify your customer to the lowest level, either by name or community.	Why do they need this in the first place? Ask them "If I didn't give you this what would happen?" If the final customer is someone else consider their requirements as well. How many suppliers are in the chain? Can you give your product or service directly to the final customer? Can the chain be reduced or eliminated?	When are they happy with what you provide? Does it have to be: • Error free • On time • Understandable • In a format they can use • The same every time • Delivered to the right place to the right person • What they asked for in the first Ask your customers to give you satisfaction indicators that you can use to judge your products or services quality.	This is the *work* you do to make the product or perform the service. What skills, training, and resources do you need to do your job? Can you identify any shortfalls or bottlenecks that prevent you from delivering the product or service the way the customer wants? Your activity might include: analysis, research, coordination, preparing reports or briefings, building data bases, organizing, managing, controlling.

FIGURE 12.1   The COPIS worksheet.

INSTRUCTIONS

1. Begin the identification process by starting with item 1, (Products and Services).
2. Work right keeping relationships between blocks.

6. PRODUCTS AND SERVICES YOU NEED TO DO YOUR VALUE ADDED ACTIVITY	7. SUPPLIERS	8. KEY SUPPLIER QUALITY REQUIREMENTS / INDICATORS	9. KEY SUPPLIER SATISFACTION INDICATORS	10. KEY DATA SYSTEMS, RELATED QPMS OR METRICS
These are the things that you need to do the work. They are inputs. Information, data, guidance, policy, regulations, reports, coordination, approval, defined requirement.	Who delivers the supplies you need to accomplish the work? It is an agency? Can you pinpoint the office or person?	What supplies do *you* need to accomplish the work? What are *your* requirements? How fast, how much, in what format, how do you want it delivered?	When are *you* happy with the products and services you need? this is where you form partnerships with your supplier to satisfy a mutual customer. *Your customers* become your *supplier's customers*, so your suppliers need to know the customer requirements as well. Set up feedback channels to reinforce their efforts when they meet your requirements and give them vital information when they don't.	What data or information do you have that tells you meet or exceed customer requirements? What is important about this process from the customers viewpoint? If it is filing a request quickly then look at cycle time. If it is an error free product, look at defect rate. Look at the customer and supplier satisfaction indicators for possible metrics.

FIGURE 12.1 (continued)

- Complete the COPIS worksheet Steps 1 through 9
- Piece together what you have so far
- Tweak your measures and environmental performance evaluators — Step 10
- Tie this information into your EMS manual or strategic plan

COPIS Step 1. Determine the Key Products and Services You Provide

You generate output in the form of products and services. In this step, identify all of your products and services you provide from the particular process you are analyzing. The best approach to this might be a brainstorming session at the section level with all members that are involved with the process. Ask yourselves, "What things (products and services) do I (we) provide to my (our) customers?" In an office environment, products and services could include information, advice, analysis, estimates, plans, documents of all types, research results, and recommended courses of action. List each product or service separately on the COPIS worksheet.

Before moving on, here is a technique you can use to define your key processes. List each product and service on a 3 × 5 card. Use an affinity diagram to sort it all out. Use the following topic sentence: What are the key processes that produce these products and services? You may have to make duplicate products and services, as some processes produce several outcomes. The headers become your key processes.

COPIS Step 2. Identify Your Customers

You give your products and services to someone — who is it? In this step you will identify all of your customers for each and all of your products and services. On the COPIS worksheet, list the customer by name that directly receives your product or service. Success in this step relies on your identifying your customer as a person. Another organization is not a customer. List the customer by name, or at least by the function, i.e., the director of the regional air quality office.

COPIS Step 3. Identify Your Customer Requirements

If you buy something, do you expect it to work? If it doesn't work you probably return it and exchange it for a working model or get your money back. What about your customers? What do they expect out of your products and services? In this step you need to ask your customers pointblank exactly what they expect from you. You may have to ask leading questions like when, how much, what, and even why. Try to get the best definition of the requirement as you can. Understanding your customer requirements may allow you to change your processes that generate waste or otherwise improve your processes.

COPIS Step 4. Identify Your Customer Satisfaction Indicators

Once you have the requirements — and both you and your customer agree to them — ask your customer this: When do you know you are satisfied with a product or

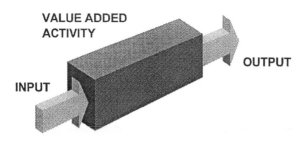

FIGURE 12.2 The value-added activity.

service I provide you? When you know the answer to that, ask them, "What are the key indicators you look for in my products and services that satisfy your needs?" Answer this right and you have your first measures — customer satisfaction and defect rate. By meeting these indicators you should reduce your waste-generation rates as well.

COPIS Step 5. Your Value-Added Activity

This next step leads you to identifying suppliers and can be used later in developing metrics and analyzing processes. Earlier you identified all your products and services, and the customers for each. Now for each product or service, you need to identify your value-added activity (Figure 12.2). What is a value-added activity? A value-added activity is that step in the process that allows you to create the products and services you provide. It probably is also the step where the most waste is generated or where the most inefficiencies exist. For example, if your product is a budget estimate, your value-added activity or activities might include research, financial analysis, and program review. To do your value-added activities you require training, skills, and knowledge relating to the work you are doing.

COPIS Step 6. Products and Services You Need to do Your Value-Added Activity

You might require information, services, or other products from other sources for you to do your work — these are inputs to your process. At this point you should identify your supplies and materials — this is a key step, because in many processes, waste generation is directly related to the quality of the product supplied to you. Identify the products and materials supplied to you; do they meet your expectations? Do they meet your customer's expectations? Can the supplies be changed to a less toxic or hazardous material while still meeting the customer requirements?

COPIS Step 7. Suppliers

Somebody has to supply you with the products and services you need to do your job — who? Try to identify all the suppliers into your value-added activity, by first identifying all the inputs you need to do your job. Next, identify the source of each input — the suppliers. List all your suppliers for each product and service. Can you replace with less toxic materials and still produce products that meet customer requirements?

COPIS Steps 8 and 9. Key Supplier Performance Indicators

Have you ever talked to your suppliers about what you expect of their products and services? Do your suppliers deliver on time, the right thing? Do you get things you don't need from your suppliers? Pick up the phone or E-mail each supplier and address your needs in specific terms. Explain to them what your customers expect and how their products and services affect the whole process. Now list what things about your suppliers' products and services satisfy you. Don't forget in this step the products and services of your supervisor. What training do you need to do your job or meet EMS requirements? What support do you need from the boss to do what you do? What policies, priorities, and guidance do you require to do your job? When was the last time you covered expectations with your supervisor?

When you finish these steps, you should have a list of requirements for each product or service you need. For each requirement, list indicators that tell you if you are satisfied. Some examples are: Is the product or service on time? Is it in the right format? Is it current? Do you have the right amount? Finally, is it a finished product?

"Time-Out": Piece Together What You Have

Before moving on to the final step of determining performance measures, it's best to piece together what you have and see how it all fits. You now have all the elements to define your processes. You should be able to graphically define your processes. Start by defining the inputs, work activity, and outputs.

Next add in the customer requirements that started the whole process in the first place. Finally, add all the key pieces and details you can possibly think of. A flowchart would be a good tool to use for analysis. Ask yourself the following questions.

- What is the customer's requirement?
- Do I understand the requirement?
- What do I need to satisfy the requirement?
- Who can supply the things I need to satisfy this requirement?
- How does my supplier deliver the products and services I need?
- Did I get what I needed from my suppliers?
- What do I do to create the final product?
- Have I identified how the process impacts the environment?
- Have I identified areas that produce waste?

Flow chart out these few simple steps. This is only a preliminary process definition. You will need to go back into the flow chart and add in details. The details are all the barriers you have to overcome. Your value-added activity needs to be exploded out in detail as well. Think about the delivery of products and services

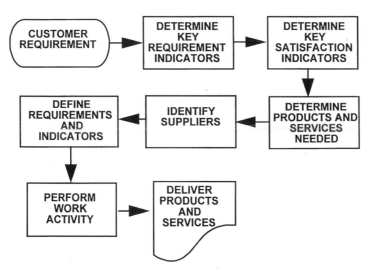

FIGURE 12.3   The COPIS flowchart

from both your suppliers and to your customers — how is that done? This step will take time. Start with the simple diagram, then add details as you recognize relationships and activities you do to meet the requirement.

Use the template (Figure 12.3) below to begin with in your flow-charting activity. Slowly start adding in detailed activities. For instance, add in coordination steps, work activities, waste generation points, regulatory inputs, permits, and don't forget the things your suppliers do.

COPIS Step 10. Measurements and Process Improvement

What do you measure? Are you measuring the right things? These are difficult questions to answer. As we discussed in the last chapter on Environmental Performance Evaluations, developing appropriate measures is a difficult thing to accomplish. Once you have completed the COPIS steps 1 to 9 you should have a better idea of what to measure. In other words, if you did all previous steps you should be ready. If not, look at your KRAs, objectives, and processes again, and focus on only a few processes first.

Determine which objectives or processes have the most influence on your reducing your process impacts on the environment. You can find out cause and effect of your KRAs on each other by using an interrelationship digraph (one of the seven management and planning tools) — but that will only show relationships of key issues. Another way to determine your impacts is to construct a simple matrix like you did in Chapter 5 on Vulnerability Analysis. First, list your top impacts (four or five should do it) along the vertical axis. Next, across the top, list out your KRAs or key processes. Then set criteria to measure with. The criteria can be: strong impact (worth 5 points), some impact (worth 3 points), and little impact (worth 1 point). Look at the totals and do some analysis. You might want to look at the aspect of "waste generation" as a key metric for everything because it has such a large impact. Your goal might be to reduce on waste generation of all key products by 25%. From here you can also identify the key performance drivers. Your objective is to find those essential performance measures that are key to your business. What are the processes, practices, and functions that are critical to satisfying customer expectations? You can poll other parts of your organization and survey customers to generate ideas for this step. There are several tools[2] you can use to narrow down the list of ideas into solid drivers:

- Affinity diagrams to identify common factors
- Interrelationship digraphs or fish-bone diagrams to relate cause and effect
- Tree diagrams constructed using the five-why method of searching for root causes

Once you determine the performance drivers, you can also construct a matrix comparing customer expectations and performance drivers. You will determine the strength of the relationship between performance drivers and customer expectations. This will provide you with the performance drivers that the customers value the most.

Correlations between customer expectations and performance drivers require the use of data or expert consensus using pareto analysis to identify the vital sources of variation in processes that affect customers' expectations or a combination of tree and matrix diagrams to show the relationships between expectations and drivers and the strength of the relationships.

To reduce impacts on the environment and facilitate process improvement, doesn't it make sense to target the things you have the most trouble with, or the things that influence your most important work? Really concentrate on these few improvement areas and you will improve move quickly than the shotgun "solve everything at once" approach. The moral of this story is: measure the KRA, objective, or process that has the most influence on your EMS.

What processes make up the objectives that have the most impact? You should have them flow-charted out by now. What are the inputs, key value-added activities, and outputs of these processes? Do you see any bottlenecks (cycle time)? Do you see any problems with suppliers (defect rate)? Does the value-added activity take too much time and resources (cycle time, cost/unit)? Are your customers satisfied with the results (customer satisfaction, defect rate)? Do the outputs always look the same (variability)?

Start simple and pick the most obvious part of the process you want to improve. Decide upon easy data collection methods. Remember that the idea is to facilitate continuous improvement, so you don't get any extra points for color and sound effects. Pick a method you can use that suits your needs.

If you're still not sure what to measure, answer this: "What are the major products and services we provide?" Now ask yourself "What are the impacts on the environment of these products and services?" You might want to start with a measure on the level of performance of a major process. You can measure the pulse and temperature of your process. What goals can you set for your process to measure from?

After you have a good feel for the process, look for trouble spots or poor performance. Why did the system perform poorly? List all the possible causes and investigate and validate them. Use a pareto chart to determine which 20% is causing 80% of the problems. Now place metrics on the top 20% troublemakers. The result will be a measure or EPE with "nested" metrics. Once you clear up the problems you don't need the "nested" metrics. Now go after the remaining problems in a similar manner.

PUTTING COPIS INTO ACTION

Steps 1 to 10 will get you to identify all the parts of your processes and how they all interact. However, the real benefit comes from you setting goals and objectives to reach for and then taking action. You can identify all the parts of a process, know how they all work and affect each other, and measure until you can't stand it any more, and the most you will accomplish is a status quo. It's like people trying to lose weight: they know which foods have more calories than others; they consistently identify all the food they eat; they measure food intake; they know who their suppliers are; and they even weigh themselves daily. They will never lose any weight — why? There is no goal or action — lots of measurement, but no action! They must first set a goal, then determine what actions they need to take to get there, then take action. The measurements help monitor, control, and speed along the process.

What are your goals and objectives at the process level? A good place to start looking is up the chain to see what goals are in the EMS manual. Those "upper level" goals are steering. What do you do that might support a particular goal or objective?

Pick among four to six of the most important things you do and concentrate on them. Write objectives that you can realistically accomplish in one year. However, don't make it so simple that you won't grow. Shoot for improvement, try stretch goals — a goal that requires you to stretch your own capabilities, but don't break them. Write out your goals and constantly review them. Remember, a goal without a plan is a dream. So carefully plan out how you will achieve your goals, what resources you need, and how long it will take. You might want to even consider using project management techniques to help you track your progress. These become the strategy portion of your EMS as we discussed in Chapter 9.

COPIS AND EMS PLANNING

Doing the COPIS tasks discussed here are only a single step. If you only do this process once, you will stumble and fall later. Set up a process where you periodically revisit steps 1 through 10. Remember, the EMS planning process will be cyclic; plug into that cycle and begin your own improvement cycle. Here is a suggested process plan:

- Accomplish COPIS steps 1 through 10,
- Compare to the goals and objectives in the EMS or strategic plan. Is there a link? Redefine as required.

- Define action plan/identify resources to meet objectives; these become the strategy portion of your EMS manual as discussed in Chapter 11. "Plug" this into your strategic plan.
- Every 6 months review COPIS.
  - Do they still make sense?
  - Has anything changed?
- Every 3 months, check progress of objectives against the environmental performance evaluation measures and other process measures and outcome measures.
  - Do they still make sense?
  - Are the dates and resources still reasonable?
  - Did anything change that will affect your plan?
- Continue to monitor and measure your processes.
- After 10 months, use the self-assessment checklists in Chapter 11 to determine your EMS health.
- After 11 months, complete an overview of the year's activity and progress, redo COPIS, in preparation for an annual EMS review or strategic planning session

BENCHMARKING FOR PROCESS IMPROVEMENT

Let's suppose you have a process that needs significant improvement in order to reduce your impacts to the environment or to meet your P2 reduction goals. By now you will have completed COPIS, and flowcharted the process in great detail so you fully understand your customer requirements, and understand how you impact the environment. It may be time to consider business process benchmarking in order to radically improve the procedures and equipment you're using to develop the products your customers require.

Benchmarking can be described as the method of measuring products, services, and practices against the toughest competitors or those known as leaders in their field.[3] Then you learn from the measurements and adapt those better business practices into your processes to improve them. Benchmarking is not copying or stealing ideas — that's been tried before; what you end up with is an excellent copy of a process that doesn't work. The idea behind benchmarking is to adapt best practices into your processes to make them better.

To gain the best benefit from benchmarking, you must take a structured approach and do your homework before you start. The basic structure is very systematic, following the same basic steps that were developed for EMS implementation and P2 continuous improvement. The steps for benchmarking are:

- Determine where you are.
- Figure out where you want to be.
- Find somebody who has reached that point.
- Understand their process.
- Adapt their process to yours.
- Measure the improvement to determine if you've reached your objective.

This is a practical approach. There are four major steps in this benchmarking process; planning, analysis, integration, and action. Let's go through each in detail.

GETTING STARTED

The basic steps for this stage are:

- Understanding your operations and processes
- Selecting the right process and understanding it
- Determining whom to benchmark against

The first step in any improvement process is to know your own operations and processes. You need to assess how your processes work and understand the strengths and weaknesses of each. You need to break out in detail who your customers and suppliers are, what products and services you provide, and how they all interrelate. Our discussions on COPIS above and earlier discussions on strategic planning are critical to achieving this step. Also consider using a flow chart in this stage for better understanding and analysis of the processes you're studying.

The hard part is determining what to benchmark and selecting the right process. We've had numerous discussions on how to prioritize your processes throughout this text. I recommend that you go back and review these discussions prior to beginning your benchmarking journey. Here are some questions you can apply which may help you to narrow down the choice.

- What is the most critical factor to my organization's success in meeting our stated policy?
- What things are causing us the most trouble?
- What are our most valuable products and services?
- What are the things that satisfy my customers?
- What specific problems has the organization as a whole identified?
- What are the most pressing organizational priorities right now?
- What activities have the greatest impact on performance?

Once you select a process to benchmark you will need measurements to determine performance. The only way to gauge improvement is with measurements. You need to decide what's important about this process that you want to improve. Use the discussion above in COPIS and the discussion of environmental performance evaluators (EPEs) to help you in this stage. Then, answer the following questions to determine if you have an appropriate measure:

- What do my customers require?
- What is the outcome of this process?
- What am I trying to achieve with this process?
- What resources do I need to do this process?
- What are the constraints in performing this process?
- What does the "best-in-class" measure?

Measure things like defect rate, waste generation rate, cost per unit, and variability. Keep your measures simple and easy to understand. Try not to have more than one or two measures for a process unless it's extremely complicated.

DATA GATHERING

The basic structure of this stage is:

- Determine how your process works.
- Search for all available information on the process. (What information is available?)
- Determine how "they" do it.

In this step you select a candidate to become a benchmark partner. We say partner, because the process is a two-way street. In benchmarking it's not all take, you give a little in return in every process. The possibility exists that you may be able to help the candidate improve the process by benchmarking — and you should be willing and prepared to do that. You want to make sure that you go after the "best-in-class" and not just the "best-on-base," because you will invest much time and energy in this process and you want the best payoff — so take your time and do it right.

Sometimes a highly advertised "best-in-class" does not perform as advertised, so it pays to investigate the process (search for all available information on the process). Some companies or organizations, like some sports teams, suffer from the "halo effect," a condition of preconceived excellence because of previous sustained superior performance. Everyone assumes present performance is superior as well. Check out the facts first, before you spend time and energy on a possible dead end.

You will need to determine the actual candidate process performance. (How do they do it?) To do this you need outcome measurement data, customer survey data (if there is any), and flowcharts of the process. This is where you search for the enabler or enablers that make their process so good — that's what you go after. Chances are their process is nearly identical to yours, but one thing makes it better: that's the enabler — that's what you want to learn about to adapt to your process.

After your final selection, compare their process to yours, using like metrics, then determine the gap. The gap is how far you have to go to have equal results from your process. The gap is not your goal. Your goal should be to become the "best-in-class," which is better than your benchmark partner.

PLAN THE IMPROVEMENT

This is the structure of this stage:

- Decide how to change
- Tell people and train them
- Plan on it

Once you identify the process to adapt, decide the best method to integrate it into your process (decide how to change). Do you do a little at a time or reengineer the entire process? Only you can decide what's best for your people. Next, you need to tell people about the change, what the expected results will be, and train people on the new procedures. Finally, the best approach to change is developing an action plan on what you will do, what you expect from the change, and what resources you need (plan on it). Develop a time line, establish supporting measures, and define process details so everyone knows what to do. The action plan then becomes your leverage to gain the support of management. You need support from upper levels if you want to be successful. There is no better way to gain management support than to show them ahead of time the facts and figures expected from the improvement. As you can see, these steps tie directly in with numerous parts of ISO 14000 EMS development, especially training, communicating, planning, and measuring. My point is that the EMS development and implementation are not mutually exclusive from process improvement activities or day-to-day business operating activities. It's an integral part of how you manage your processes.

IMPLEMENT THE IMPROVEMENT

The structure of this stage is:

- Implement the change
- Measure and track it
- Analyze it
- Report it
- Correct it (if needed)

The next step is taking action. Once you start there may seem to be an initial decrease in performance until people get used to the new process. If you did your homework right, however, performance should start to increase. You need to track the new process by measuring the performance. Look at the results carefully and examine the whole process to ensure you're doing what you planned. Tell everybody about the performance of the new process. Nothing will kill a good plan faster than the lack of feedback; remember communication is a key part of an effective EMS.

Finally, be aware that you may have to fine-tune the new process after implementation. Someone may misunderstand the plan or might not fully know what to do so may tweak it as necessary.

FOLLOW-UP

Taking action is not the final step. The final step is to go back to step 1, planning, and analyze the process and see if any additional improvements are necessary; if so, you continue through all the stages again. Follow-up is an integral part of continuous improvement.

If properly implemented, benchmarking can yield dramatic improvements to manufacturing and business processes. But it takes time and other resources to understand your processes, followed by a very extensive data search to determine against whom to benchmark. You can reduce your data search expenses if you know where to begin. This is where the next section begins.

WHERE DO I LOOK FOR MORE INFORMATION?

As I have discussed numerous times throughout this text, you can't implement an effective EMS by yourself; you must enlist the help of others. This includes not only individuals, but also organizations external to yours. So where do you begin? There are many publications, articles, case studies, and other documents written on specific pollution prevention opportunities. These are available from the EPA as well as your local state regulatory agencies. Although most state environmental agencies have been hit hard by budget cuts, many still have active P2 programs and may be able to help you implement your programs. At the very least, they should be able to provide lists of contacts and references for you to begin searching against. The EPA can also provide numerous opportunities to learn from other organizations. They have case studies, hotlines, and how-to manuals to help you get started or continue your journey.

One of the single most rich resources of environmental management information is at your fingertips. It's on the internet. Surfing the net will identify numerous sites for further browsing and will provide you with a huge database of information with very little time invested. I've listed a few addresses below to begin your search. While this is only a partial list, these have links to many more.[4] I recommend that you call the hotlines and visit the web pages from the organizations below as you begin your data-gathering phase to find a better way to operate your processes while focusing on reducing pollution and your impact on the environment.

POLLUTION PREVENTION

1. Defense Envr. Network & Information eXchange (DENIX) Access. This is a DoD Environmental Bulletin Board, P2 technical library, and other resources. Visit on the internet/world wide web address:
   http://denix.cecer.army.mil/denix/denix.html
2. Department of Energy's Pollution Prevention Information Clearing House (EPIC) home page. This is a great bulletin board with many links to other bulletin boards. A great place to start any searches. Internet/world wide web address:
   http://146.138.5.107/EPIC.htm
3. Navy CFC/Halon Information Clearinghouse. Call (703) 769-1883 for more information about reducing or eliminating ozone-depleting chemicals.
4. Enviroene. An EPA Bulletin Board which attempts to provide a single repository for P2 information. Great place to begin a search. Internet address:
   http://es.inel.gov

5. Waste Not. A DOE bulletin board operated by Lockheed Martin Idaho Technologies. Visit at:
   http://wastenot.inel.gov
6. For a great list of other P2 web sites, visit a web site that belongs to the Environmental and Society Group of the Battelle Seattle Research Center. It is a great place to find exciting links to other sites and also has some wonderful papers to review. See
   http://www.seattle.battelle.org/services/e&s/moresite.htm
7. Alternative Treatment Technology Information Center (ATTIC). Another EPA web site which provides a comprehensive database on innovative alternate technologies. Visit at:
   http://www.epa.gov/attic
8. Center for Waste Reduction Technologies. The EBB is operated by the University of Florida Chemical Engineering Department. Visit at:
   http://www.che.ufl.edu
9. Hazardous Waste Research and Information Center (HWRIC):
   http://www.hazard.uiuc.edu/hwric/hmtlhome.html or call (217) 333-8940.
10. Point Source Information and Provision Exchange Systems (PIPES). Visit:
    http://PIPES.eshg.saic.com/PIPES.htm or call (703) 821-4697.
11. Pollution Prevention Information Clearing House (PPIC) E-mail:
    ppic@epamail.epa.gov or call (202) 260-1023.
12. Waste Reduction Resource Center for the Southeast:
    http://owr.ehnr.state.nc.us/wrrcl/htm or call (800) 476-8686.
14. SAGE — RTI's Solvent Alternatives Guide. Visit:
    http://clean.rti.org/
15. Pollution Prevention Research Projects Database gopher:
    //gopher.pnl.gov:2070/1/.pprc
16. Obtain P2TECH info from liebl@wisplan.uwex.edu archives at:
    http://gopher.great-lakes.net:2200/1s/glin/majordomo/p2tech
17. Obtain P2 info from P2_info@pnl.gov
18. This home page is a great place to begin searches and review publications and articles: http://www.envirolink.org
19. Public Information Center (PIC): Call at (202) 260-2080 or visit at:
    http://www.epa.gov/pic.html
20. Los Alamos National Lab's P2 home page. This is a great web site for reviewing P2 publications, fact sheets, and material substitution lists. Visit at:
    http://perseus.lanl.gov
21. Handbook of Environmental Evaluations. This document was developed by the Australian Department of the Environment for evaluating an organization's impact on the environment. Visit this site at:
    http://www.erin.gov.au/portfolio/esd/handbook/chamanag.html

HAZARDOUS MATERIAL INFORMATION

1. DOT Hazardous Material (800) 752-6367.
2. OSHA (800) 321-6742.

ISO 14000 REFERENCES

1. American National Standards Institute (ANSI), U.S. Delegate to ISO Technical Commitee 207. Call (212) 642-4900.
2. American Society for Testing and Materials (ASTM), Administrator for the U.S. TAG. Call (610) 832-9500.
3. ISO 14000 on the net. Begin your search at some of these web sites.
   • The ISO 14000 home page operated by ISO. Visit at: http//www.ISO14000.com
   • For great articles and links see: http//www.dep.state.pa.us/deputate/pollprev/iso14000
   • On Stoller visit the ISO 14000 home page at: http://www.stoller.com/iso.htm

OTHER RELATED TOPICS AND MISCELLANEOUS ADDRESSES

1. Environmental Financing Information Network (EFIN). Visit: http://www.epa.gov.efinpage
2. Inform: Inform@igc.apc.org
3. Infoterra/USA; visit: http://www.epa.gov/contacts/INFOTERRA.html
4. Right-to-Know Act on the net. This site provides online access to databases and other information. Visit at: http://rtk.net or call (202) 234-8494.

DESIGN FOR ENVIRONMENT (DfE) INFORMATION

1. Carnegie Mellon University Green Design Initiative Home Page: http://www.ce.cmu.edu:8000/GDI/
2. CERES Global Knowledge Network: http://www.cerc.wvu.edu/ceres/ceres_index.html
3. Pacific Northwest Laboratory's Design for Environment Page: http://w3.pnl.gov:2080/DFE/home.html
4. Sources of Environmentally Responsible Wood Products: http://www.ran.org/ran/ran_campaigns/wood_con/wood_sources.html
5. UC Berkeley Center for Green Design and Manufacturing: http://www.me.berkeley.edu/green/cgdm.html
6. University of Windsor Environmentally Conscious Design and Manufacturing Information database: http://ie.uwindsor.ca/ecdm_info.html

Recycling Information

1. Global Recycling Network:
   http://grn.com/grn/
2. Jay Stimmel's Recycling List:
   http://perseus.lanl.gov/NON-RESTRICTED/Recycle_List.html
3. King County Recycled Procurement Program:
   http://homer.metrokc.gov/oppis/recyclea.html
4. National Materials Exchange Network:
   http://www.earthcycle.com/g/p/earthcycle//
5. Pacific Northwest Laboratory's Guide to Buying Green:
   http://www.pnl.gov:2080/esp/greenguide/
6. Recycler's World: http://granite.sentex.net:80/recycle/
7. Wastewi$e: http://www.epa.gov/epaoswer/wastewise.html
8. Waste Reduction Center for the South East. Call at (800) 476-8686 or visit the site at:
   http://owr.ehnr.state.nc.us/wrrcl/html

State Pollution Prevention Programs

1. California EPA's P2 home page. Lots of good information, including fact sheets, newsletters, and links to archives. Visit at:
   http://www.calepa.cahwnet.gov.
2. Center for Neighborhood Technology. This home page supports sustainable urban communities and as such provides insight on P2 activities in urban settings. Visit at:
   http://www.cnt.org
3. Environment Canada's P2 home page. Visit at:
   http://www.ns.doe.ca/epb/progs/pollprev.html
4. EPA Office of Air Pollution Prevention Programs (Green Star, etc.):
   http://www.epa.gov/docs/GCDOAR/OAR-APPD.html
5. Industrial Opportunity Assessment Database at Rutgers University:
   http://128.6.70.23/
6. National Pollution Prevention Center for Higher Education:
   http://www.umich.edu/~nppcpub/nppc.html
7. Washington Toxics Coalition:
   http://www.scn.org/scripts/menus/w/wtc/wtc.menu

Regulatory and Policy Information

1. Cornell University Archive of Federal Regulations:
   http://www.law.cornell.edu/topics/environmental.html
2. EPA Web Site: http://www.epa.gov/
3. EPA hotlines
   - EPA Publications: (202) 260-7751
   - EPCRA/SARA Hotline: (800) 535-0202

- TSCA Hotline: (202) 554-1404
- RCRA/Superfund Hotline: (800) 424-9346
- Asbestos Hotline: (800) 368-5888
- Pesticides Hotline: (800) 858-7378
- Stormwater Hotline: (202) 260-7786
- Stratospheric Ozone Protection Hotline: (800) 296-1996 or http://www.epa.gov/ozone
- Air Hotline: (919) 541-0800
- Radon Hotline. Call (800) 767-7236 or visit: http://www.epa.gov/iaq/radon/sosradon.html.
- EPA Green Lights program. Call (202) 775-6650 or visit: http://www.epa.gov/docs/GCDOAR/greenlights.html
- RCRA, Superfund & EPCRA Hotline: (800) 424-9346
- Safe Drinking Water Hotline: (800) 426-4791
- WasteWi$e. Call (800) EPA Wise, or visit: http://www.epa.gov/epaoswer/wastewise.html

4. Internet Virtual Library — Environmental Law: http://www.law.indiana.edu/law/intenvlaw.html
5. EPA Future Studies Gopher: Gopher://futures.wic.epa.gov
6. Delaware Department of Natural Resources and Environmental Control: http://www.dnrec.state.de.us/
7. Florida Department of Environmental Protection: http://www.dep.state.fl.us/index.html
8. Georgia DNR Technical Assistance Division: http://www.state.ga.us/Departments/DNR/P2AD/
9. Illinois HWRIC: http://www.inhs.uiuc.edu/hwric/hmlhome.html
10. Ohio EPA Office of Pollution Prevention: http://www.epa.ohio.gov/opp/oppmain.html
11. Pennsylvania DEP — P2 and Compliance Assistance: http://www.dep.state.pa.us/dep/deputate/pollprev/default.htm
12. Washington Department of Ecology Home Page: http://olympus.dis.wa.gov/www/access/ecology/ecyhome.html
13. Maine Department of Environmental Protection's P2 Resource List: http://www.state.me.us/dep/p2list.htm

TRADE AND PROFESSIONAL ASSOCIATIONS

1. Air and Waste Management Association: http://www.awma.org/
2. Environmental Industry's Home Page: http://www.cleanup.com/welcome.html
3. National Association of Environmental Professionals: http://www.enfo.com:80/NAEP/Main/
4. National Center on Manufacturing Sciences: http://www.ncms.org/

5. National Oil Recycler's Association:
   http://www.webcom.com/~infoserv/customer/nora/welcome.html
6. Polymers Home Page:
   http://www.polymers.com/
7. Environmental Professional's Home Page:
   http://www.crl.com/~grega/envpage.html#1
8. American Institute for Pollution Prevention (AIPP) — a trade association
   for P2 professionals. Call (202) 797-6567, or visit:
   http://es.inel.gov/aipp
9. Solid Waste Association of North America (SWANA). Visit at:
   http://www.swana.org.

POLLUTION PREVENTION SOFTWARE

There are a few software programs available to assist you with your P2 activities. Scott Butner from the Environment and Society Group of the Battelle Seattle Research Center has identified many of these,[5] which are included in the discussion to follow. Since pollution prevention encompasses such a broad range of activities, from good housekeeping and improved management practices to product and process redesign, it can be a bit difficult to determine just exactly what qualifies as "pollution prevention" software. Some of the available software is included here. Others can be found via the internet.

EarthAware. This is an effective environmental education tool that emphasizes a preventive approach to the environmental choices we each face in our daily lives. Using a combination of hypertext tutorials, databases, and links to internet sites, EarthAware is an attractive piece of software that may provide even a veteran pollution prevention planner with new ideas for reducing waste. EnviroAccount Software (1-800-554-0317 or 916-756-9156)

Solvent Alternatives Guide (SAGE). SAGE provide a technically competent, user-friendly source of information on cleaning technologies. SAGE is an "expert system" that leads you through a series of questions that help the system narrow down the cleaning options based on wide variety of part and process specific issues. You're then presented with a detailed technical description of each applicable option, along with the rationale for its selection. Perhaps the most valuable way to use SAGE is to pay attention to the types of questions it asks in selecting technologies. This is a great way to sharpen your thought processes in making cleaning technology decisions, and the technology descriptions contained in SAGE are often as detailed and to the point as anything else available. For more information, contact Research Triangle Institute (919-541-6916).

Pollution Prevention Electronic Design Guide (P2EDGE). P2EDGE is essentially an electronic "idea notebook" of pollution prevention strategies that can be applied to the design and construction of buildings and other fixed facilities. Originally designed for the U.S. Department of Energy for use by their facility planners, the P2EDGE software is used in conjunction with the Pollution Prevention Design

Opportunity Assessment protocol to help designers identify opportunities for minimizing the impact of hazardous material spills, reducing hazardous material inventories, reducing stormwater pollution, and incorporating recycled materials into construction of buildings, parking lots, and landscaping. Many of the strategies are accompanied by graphic examples, bibliographic citations, or other supplemental information. A checklist lets you keep track of which ideas are being evaluated for inclusion in the building design, who's responsible, and other information related to implementation. P2EDGE is a fun tool to use for inspiration if you are involved in building, upgrading, or operating a facility. New versions are currently under development that will expand the P2EDGE database to include pollution prevention strategies in chemical process design and in textile product design. The software requires Windows and works best on a 486-class machine or better. Contact the Pacific Northwest National Laboratory (509-375-3703) for more information.

SOLUTIONS Facilities P2 Plan. SOLUTIONS Software specializes in providing public-domain documents (regulatory lists, manuals, etc.) on CD-ROM. Their Facilities Pollution Prevention Plan CD-ROM (compatible with Windows) includes complete texts of federal pollution prevention regulations, executive orders, and guidance documents, along with a hypertext version of the EPA's Facility Pollution Prevention Planning Guide. This tool can be especially helpful to small businesses seeking help in organizing their pollution prevention plans, and provides a rich source of "boilerplate" language for plans, permitting the user to focus on putting their creative energy into identifying and implementing pollution prevention, rather than writing about it. Contact SOLUTIONS Software Corporation (407-321-7912).

RLIBY. RLIBY is technically not a software tool, but rather a database of pollution prevention articles, pamphlets, and other documents maintained by the Waste Reduction Resource Center in Raleigh, NC, arguably one of the largest pollution prevention clearinghouses in the U.S. The WRRC's library contains more than 6000 entries, covering most industry sectors. The database was created for use with PROCITE software; a read-only version of the software is available (for DOS) with the database upon request from WRRC. Contact the Waste Reduction Resource Center (800-476-8686).

Waste Reduction Advisory System (WRAS). WRAS is one of the oldest software packages for pollution prevention, developed for the state of Illinois in 1987. It is basically a MSDOS-compatible database of pollution prevention and waste minimization article abstracts, organized by both keyword and Standard Industrial Code (SIC). While the articles are somewhat dated, the simplicity of the tool, and the quick access to pollution prevention strategies make it a very good program to have. Though HWRIC has no current plans to maintain the software, the low price ($20) makes it a worthwhile investment. Contact the State of Illinois Hazardous Waste Research and Information Center.

P2/FINANCE. Originally developed for the U.S. EPA by the Tellus Institute as a means of illustrating environmental cost accounting principles, particularly for the evaluation of capital investment options such as process upgrades or modifications. The software helps users identify the full range of environmental costs associated with a process, by taking into account such factors as waste management and liability costs, which often get "lumped" into overhead accounts. The original version, still supported by Tellus, was developed as a spreadsheet template, with versions available for Microsoft Excel or Lotus 1-2-3. Industry-specific versions of P2/FINANCE are being developed by Tellus for the screen and lithographic printing industries, printed wire board fabrication, and metal fabrication industries. Considerable effort has gone into the update of the original spreadsheets, and the current versions of P2/FINANCE are implemented as stand-alone FoxPro applications.

GREENWARE Environmental Systems Inc. This company has developed three software products based on ISO 14000. The first product is used for EMS self assessments, the second helps with implementation, while the third is an audit software package. For more information see: http://www.greenware.ca/software/iso14000.htm

In addition to the tools discussed above, there are many other software tools that can be useful in implementing a pollution prevention program. For example, a quick check of the internet will reveal dozens of commercial MSDS databases, hazardous material tracking systems, and waste tracking software products that can be essential tools in assessing pollution prevention opportunities and measuring program effectiveness.

CLOSING THOUGHTS

The EMS and pollution prevention journey can only be successful if it's continuously focused upon and continuously improved. When you first begin your journey, you'll want to keep it simple, but a time will come when you move beyond the simple implementation to more complex process modifications. This will require a more detailed understanding of your processes, procedures, and tasks. It will also require more focused and advanced decision-making tools and greater research on alternate methods of meeting your customer requirements. By following the information provided within this chapter and throughout the text you will be prepared for the challenges you'll meet on your journey.

REFERENCES AND NOTES

1. Caldwell, R., Process Improvement Guide, U.S. Air Force Quality Improvement Office publication, Langley AFB, VA. An internal USAF document, undated.
2. For a detailed discussion of the tools I mentioned within this section, I recommend that you read: *The Memory Jogger Plus; The Seven Management and Planning Tools,* by Michael Brassard, Goal/QPC, Methuen, MA, 1989.
3. Information in this section was adapted from a U.S. Air Force document, Benchmarking, written by the USAF, Quality Office, Langley AFB, VA. An internal USAF document, 1994. Much has been written on benchmarking business processes. I recommend that you review two books if you want more information: a. Camp, R.C., Benchmarking, in *The Search for Industry Best Practices That Lead to Superior Performance,* ASQC Quality Press, Milwaukee, 1989. b. Watson, G.H., *The Benchmarking Workbook,* Productivity Press, Portland, OR, 1992.

4. The information on this list was modified with permission from two sources: a. Butner, S., *Pollution Prevention Links on the Internet,* Battelle's On-Line Pollution Prevention Library, August 1995. b. Blevin, T., Environmental Resource Handbook, U.S. Air Force, Air Combat Command, Langley AFB, VA, 1995. This document was developed for the Air Force by Radian Corporation, Alexandria, VA.

5. Modified with permission from: Butner, S. *Software Tools for Pollution Prevention,* from Battelle's on-line pollution prevention library, originally prepared for Northwest Pollution Prevention Resource Center "Pollution Prevention Northwest," last updated July 1996.

GLOSSARY OF TERMS

Affinity Diagram: A brainstorming and management tool that assists with general planning. It makes disparate language information understandable by placing it on cards and grouping the cards together in a creative manner. "Header" cards are used to summarize each group.

Affirmative Procurement: The purchase of environmentally preferable products, required of Federal agencies by RCRA Section 6002 and EO 12783. Affirmative procurement programs must establish preference for products containing recycled material, must include a promotion plan to place emphasis on buying recycled, and must have procedures for obtaining and verifying estimates and certifications of recycled content.

Alternatives: Ways of reducing adverse effects of hazardous materials. Alternatives, as applied to hazardous material decision-making, include, but are not limited to, such possibilities as substituting less hazardous or nonhazardous material; redesigning a component such that hazardous material is not needed in its manufacture, use, or maintenance; modifying processes or procedures; restricting users; consumptive use; on-demand supply; direct ordering; extending shelf life; regenerating spent material; downgrading and reuse of spent material; use of waste as raw material in other manufacturing and combinations of those factors.

Alignment: The process of improving a system so that all elements contribute to the aim, such as the EMS goals aligning with other organization goals.

ANSI: American National Standards Institute.

ASQC: American Society for Quality Control.

Assessment: An evaluation process used to measure the performance or effectiveness of a system.

ASTM: American Society for Testing and Materials.

Audit: As related to auditing an EMS. A planned, independent, and well documented assessment process to determine whether agreed-upon requirements are in conformance with the EMS and the EMS standards. It's usually conducted as an independent examination.

Auditor: A individual qualified to perform audits.

Baseline: Quantified starting points from which progress is measured. A beginning point based on an evaluation of the output over a period of time to determine how the process performs prior to any improvement effort. Pollution prevention baselines are usually shown as quantities of material purchased or waste generated over a specified period of time.

Benchmarking: The process of finding and adapting best practices to improve organizational performance.

Best Practice: A superior method or innovative practice that contributes to improved performance.

Brainstorming: An idea-generating technique that uses group interaction to generate many ideas in a short time period. Ideas are solicited in a nonjudgmental, unrestricted manner from all members of a group.

BS 7750:   Bristish Standard 7750 for Environmental Management

Cause-and-Effect Diagram: A diagram graphically illustrating the relationship between a given outcome and all the factors that influence this outcome. Also called a fishbone diagram.

CEMP: U.S. EPA's Code of Environmental Management Practices being developed by the Office of Facilities Enforcement.

CERCLA: Comprehensive Environmental Response, Compensation, and Liability Act.

Certified: An organization is certified following an assessment by an accrediting body, indicating that the organization is meeting the requirments of an EMS standard.

Charter: A written commitment by management stating the scope of authority and the resources available for an improvement team.

Checksheet: A form for recording data on which the number of occurrences of an effect can be recorded such as tick marks or checks.

Closed-loop Recycling: Utilization of by-products from a production process in the original process, without significant alteration or reprocessing. There are three key requirements: the by-product must be returned to the process without first being reclaimed (i.e., distilled, dewatered, or treated); the production process to which the by-product is returned must be a primary production process (a process that uses raw materials as the majority of its feedstock); and the by-product must be returned as feedstock to an operation within the original process from which it was generated.

CMA: Chemical Manufacturers Association.

Compost: A mixture of garbage and degradable trash with soil in which certain bacteria in the soil break down the garbage and trash into organic fertilizer.

Composting: A waste management option involving the controlled biological decomposition of organic material in the presence of air to form a humus-like material. Controlled methods of composting include mechanical mixing and aerating, ventilating the materials by dropping them through a vertical series of aerated chambers, or placing the compost in piles out in the open air and mixing or turning them periodically.

Conformance: An indication that the product of service meets or exceeds the requirements of relevant specifications, contract, or regulation.

Continuous Improvement Process: A diagram graphically illustrating the steps followed to implement improvement activities and to ensure that improvement is on-going to increase customer satisfaction and improve organizational efficiency.

Corrective Action: Action taken to eliminate causes of an existing nonconformance or deficiency.

Cost–Benefit Analysis: A method to compare the cost and benefits of proposed plans. It can be used for comparing the financial outcomes of different actions and determining if a particular action makes sense financially.

DIS: Draft International Standard.

DoD: U.S. Department of Defense.

EAPS: Environmental Aspects of Product Standards.

EMAS: Eco-management and Audit Scheme.

EMS: Environmental Management System. The application of total quality management to environmental management.

EPA: U.S. Environmental Protection Agency.

EPE: Environmental Performance Evaluation.

EPI: Environmental Performance Indicators.

EPA-17 Targeted Chemicals: Seventeen chemicals (or compounds) selected for reduction or elimination based on their volume of use, toxicity, persistence, and mobility. Also known as EPA industrial toxic pollutant (ITP) chemicals.

EPCRA: Emergency Planning and Community Right-to-Know Act. Sometimes referred to as the SARA Title III requirements.

EU: European Union.

Five "Whys": A technique for discovering the root cause(s) of a problem and showing the realtionship of causes by repeatedly asking the question "why".

Flowchart: A graphic structured representation of all the major steps in a process.

Fluorocarbons (FCs): Any of a number of organic compounds analogous to hydrocarbons in which one or more hydrogen atoms are replaced by fluorine. Once used in the U.S. as propellants in aerosols, they are now primarily used in coolants and some industrial processes. FCs containing chlorine are called chlorofluorocarbons (CFCs). They are believed to be modifying the ozone layer in the stratosphere, thereby allowing more harmful radiation to reach the earth's surface.

Force Field Analysis: A technique that helps identify and visualize the relationships of significant forces that influence a problem or impact a goal.

Functional Areas: The operations or areas of responsibility that affect or are affected by the use of hazardous material. These areas include, but are not limited to, budget and fiscal planning, legal support, research and development, weapons systems acquisition and maintenance; material and performance specifications and standards; design handbooks and technical manuals; maintenance and repair procedures; industrial processes; procurement policy; contracting provisions; new material identification; public works operations; construction; management of munitions; chemical agents; propellants; medical and other personnel support; safety and occupational health; transportation and logistics analysis; supply; warehousing; distribution; recycling; disposal; spill prevention, control, and cleanup; contaminated site remediation; staffing, education, and training; information exchange; public affairs; general administration; and oversight.

Gap Analysis: The comparison of a current condition to the desired state.

Goal: A broad statement describing a desired future condition or achievement, without being specific about how much and when.

Halon: Bromine-containing compounds with long atmospheric lifetimes whose breakdown in the stratosphere causes depletion of ozone. Halons are used in fire-fighting.

Hazardous Material Pharmacy: Single point of control for hazardous material. The concept is that hazardous material is as valuable as a controlled pharmaceutical drug and should be treated as such. Sometimes also referred to as a "HAZMART"

Hazardous Material: Any material that poses a threat to human health and/or the environment typically due its toxic, corrosive, ignitable, explosive, or chemically reactive nature.

Hazardous Substance: 1. Any material that poses a threat to human health and/or the environment. Typical hazardous substances are toxic, corrosive, ignitable, explosive, or chemically reactive. 2. Any substance named by U.S. EPA to be reported if a designated quantity of the substance is spilled in the waters of the U.S. or if otherwise emitted into the environment.

Hazardous Waste: By-products of society that can pose a substantial or potential hazard to human health or the environment when improperly managed. Possess at least one of four characteristics (ignitability, corrosivity, reactivity, or toxicity), or are listed in 40 CFR 261.30 or applicable state or local waste management regulations.

Histogram: A chart which displays measurements as distribution indicating the amount of variation within a process.

Implementation: A structured approach that addresses all aspects (who, what, when, why, and how) of incorporating improvements into a process or system.

Improvement: The organized creation of beneficial change.

Industrial Solid Waste: Includes wastewater treatment sludges, solids from air pollution control devices, trim or scrap materials that are not recycled, fuel combustion residues (such as the ash generated by burning wood or coal), and mineral extraction residues.

Inhibitors: Individual managers and or workers unwilling to promote improvement activities regardless of demonstrated results of reasoning.

Inputs: Product or services obtained from others (suppliers) in order to perform your job tasks.

ISO: Internation Organization for Standardization. A worldwide organization which promotes standards for international trade, manufacturing, and communication. ISO is not an acronym, it comes from the Greek word "isos," meaning equal.

ISO 9000:   Quality Management Standards.

ISO 14000:   Environmental Management Standards.

Key Process: The major system level processes that support the mission and satisfy major customer requirements. The identification of key processes allows the organization to focus its resources on what is important to the customer.

Key Focus Area: The major category of customer requirements that is critical for the organization's success.

Life Cycle Assessment (LFA):   A systematic method of determining how a product impacts the environment throughout its life cycle from production, to use and disposal. This includes evaluating material and energy inputs and outputs.

Life Cycle Economic Analysis: An evaluation of the cost associated with the use of hazardous material and potential alternatives over the life of the investment or hazardous material. The analysis is not a specific, step-by-step procedure that can be applied by rote to all cases. Analysis shall be guided by basic principles of economics and informed judgment.

Management System: A structured, nontechnical system describing the policies, objectives, principles, organizational authority, responsibilities, accountability, and implementation plan of an organization for conducting work and producing items and services.

Mass Balance: An analytical method of accounting for the quantities of materials produced, consumed, used, or accumulated within a process.

Measurement: The act or process of quantitatively comparing results to requirements to arrive at an estimate of performance.

Metrics: Measurements, taken over time, used to communicate vital information about a process or activity. A metric should drive appropriate management action. The metric package should consist of an operational definition, measurement over time, and the graphical presentation.

Municipal Solid Waste (MSW): Wastes generated by administrative and domestic activities. Includes wastes such as durable goods, nondurable goods, containers and packaging, food wastes, yard wastes, and miscellaneous inorganic wastes from residential, commercial, institutional, and industrial sources. MSW does not include wastes from other sources, such as municipal sludges, combustion ash, and industrial nonhazardous process wastes that might also be disposed of in municipal waste landfills or incinerators.

Multivoting: A structured voting process used to reduce a large number of items, usually ideas, to a more manageable number for further processing or analysis.

Nominal Group Technique (NGT): A tool for generating a list of ideas or opportunities. Priorities are determined by voting or ranking.

Nonconformance: Failure to meet or fulfill a specific requirement.

NPDES: National Pollutant Discharge Elimination System.

Objectives: A specific statement of a desired shorter-term condition or achievement. Includes measurable end results to be accomplished by specific teams of people within time limits. It is the "how," "when," and "who" for achieving a goal.

Open-loop Recycling: A recycling system in which a product made from one type of material is recycled into a different type of product. The product receiving recycled material itself may or may not be recycled.

Opportunity Assessment: Systematic procedures to identify and assess ways to prevent pollution by reducing or eliminating wastes.

OSHA: Occupational Health and Safety Administration.

Outputs: Products, materials, services or information provided to customers.

Ozone (O_3): Found in two layers of the atmosphere, the stratosphere and the troposphere. In the *stratosphere* (the atmospheric layer beginning 7 to 10 miles above the earth's surface), ozone is a form of oxygen found naturally which provides a protective layer shielding the earth from ultraviolet radiation's harmful effects on humans and the environment. In the *troposphere* (the layer extending up 7 to 10 miles from the earth's surface), ozone is a chemical oxidant and major component of photochemical smog. Ozone can seriously affect the human respiratory system and is one of the most prevalent and widespread of all the criteria pollutants for which the Clean Air Act required EPA to set standards. Ozone in the troposphere is produced through complex chemical reactions of nitrogen oxides, which are among the primary pollutants emitted by combustion sources; hydrocarbons, released into the atmosphere through the combustion, handling, and processing of petroleum products; and sunlight.

Ozone Depletion: Destruction of the stratospheric ozone layer which shields the earth from ultraviolet radiation harmful to biological life. This destruction of ozone is caused by the breakdown of certain chlorine-, fluorine-, and/or bromine-containing compounds (chlorofluorocarbons or halons) which break down when they reach the stratosphere and catalytically destroy ozone molecules.

Ozone Depleting Substances (ODS) and Ozone Depleting Chemicals (ODCs):   CFCs, halons, and other substances that deplete the stratospheric ozone layer as classified by the Clean Air Act of 1990.

Pareto Chart: A statistical method of measurement to identify the most important problems through different measurements scale such as cost, or frenquency. It directs attention and efforts to the most significant problems.

PDSA: Plan-Do-Study-Act (or Plan-Do-Check-Act). A structured cyclical methodology for developing and implementing actions of any type. Also called the Deming Wheel or the Shewhart Cycle.

Policy: Overarching plan or direction for achieving organization's goals.

Pollution/Pollutants: Refers to all nonproduct outputs, irrespective of any recycling or treatment that may prevent or mitigate releases to the environment.

Pollution Prevention: An organized, comprehensive effort to systematically reduce or eliminate pollutants or contaminants prior to their generation or their release into the environment. It is the use of materials, processes, or practices that reduce or eliminate the creation of waste at the source.

Pollution Prevention Hierarchy: The Pollution Prevention Act of 1990 established a hierarchy as national policy. The hierarchy follows this order: (1) prevent or reduce pollution at the source wherever feasible; (2) recycle, in an environmentally acceptable manner, pollution that cannot feasibly be prevented; (3) treat pollution that cannot feasibly be prevented or recycled; and (4) dispose of, or otherwise release into the environment, pollution only as a last resort.

Process: A set of interrelated work activities that are characterized by a set of specific inputs and value-added tasks that produce a set of specific outputs.

Process Owner: The one person who has the authority to make changes to a process.

RCRA: Resource Conservation and Recovery Act.

Recycle/Reuse: The process of minimizing the generation of waste by recovering usable products that might otherwise become waste. Examples are the recycling of aluminum cans, waste paper, POLs, engine coolants, and ODS.

Recycled Content: The amount of recovered material, either pre- or postconsumer, in a finished product that was derived from materials diverted from the waste management system. Usually expressed as a percent by weight.

Refuse Reclamation: Conversion of solid waste into useful products, e.g., composting organic waste to make soil conditioners or separating aluminum and other metals for melting and recycling.

Registration: A formal process in which an accredited body audits a supplier's processes and verifies that a defined set of performance characteristics have been met. Registration occurs after successfully passing an ISO 14000 audit.

Return on Investment (ROI): An attempt to quantify whether the benefits to a particular project exceed costs.

SAGE: Strategic Advisory Group on the Environment.

SARA: Superfund Amendments and Reauthorization Act. Also see EPCRA.

SC: Subcommittee.

Source Reduction: Any practice that reduces the quantity of hazardous substances, contaminants, or pollutants. The Federal Pollution Prevention Act defines source reduction as "any practice which (1) reduces the amount of any hazardous substance, pollutant, or contaminant entering any waste stream or otherwise released into the environment (including fugitive emissions) prior to recycling, treatment, and disposal; and (2) reduces the hazards to public health and the environment associated with the release of such substances, pollutants, or contaminants. The term includes equipment or technology modifications, process or procedure modifications, reformulation or redesign of products, substitution of raw materials, and improvements in housekeeping, maintenance, training, or inventory control." Source reduction does not entail any form of waste management (e.g., recycling and treatment). The Act excludes from the definition of source reduction "any practice which alters the physical, chemical, or biological characteristics or volume of a hazardous substance, pollutant, or contaminant through a process or activity which itself is not integral to and necessary for the production of a product or the providing of a service."

Stakeholder: Any individual, group, or organization that will have a significant impact on, or will be significantly impacted by, the quality of the product or services you provide.

Story Board: A technique to graphically display the methodology used and progress made by an improvement team.

Strategic Planning: The process by which an organization envisions its future and develops strategies and plans to achieve that future.

Subprocess: The process that makes up a larger process.

System: A group of interdependent processes and people that together perform a common mission.

TAG: Technical Advisory Group.

Task: Specific, definable activities to perform an assigned piece of work, often finished in a certain time.

TC: Technical Committee.

TC/207: ISO 14000 Technical Committee Number 207, responsible for developing the "Environmental Management System Standards."

Toxic Chemical: Those chemicals listed in 40 CFR 372.65, also known as the Toxic Release Inventory (TRI) List. The list changes periodically as updates are published in the *Federal Register.*

Toxic Chemical Use Substitution: This term describes replacing toxic chemicals with less harmful chemicals, although relative toxicities may not be fully known. Examples would include substituting a toxic solvent in an industrial process with a chemical having lower toxicity, or reformulating a product so as to decrease the use of toxic raw materials or the generation of toxic by-products. This also includes attempts to reduce or eliminate the use of chemicals associated with health or environmental risks. Examples include the phaseout of lead in gasoline, the phaseout of the use of asbestos, and efforts to eliminate emissions of chlorofluorocarbons and halons. Some of these attempts may involve substitution of less hazardous chemicals for comparable uses, while others involve the elimination of a particular process or product from the market without direct substitution.

Toxic Use Reduction: This term refers to the activities grouped under "source reduction," where the intent is to reduce, avoid, or eliminate the use of toxics in processes and/or products so as to reduce overall risks to the health of workers, consumers, and the environment without shifting risks between workers, consumers, or parts of the environment.

Treatment: Involves end-of-pipe destruction or detoxification of wastes from various separation/concentration processes into harmless or less toxic substances.

Vulnerability Assessment: An assessment performed early in the process analysis stage for the purpose of determining which areas present the greatest risk to the organization.

Waste Minimization: Source reduction and the following types of recycling: (1) beneficial use/reuse, and (2) reclamation. Waste minimization does not include recycling activities whose uses constitute disposal and burning for energy recovery.

WG: Working Group as part of a subcommittee.

Index